Astrophysics and Space Science Library

More information about this series at http://www.springer.com/series/5664

David H. Levy

The Starlight Night

The Sky in the Writings of Shakespeare, Tennyson, and Hopkins

Second Edition

Springer

David H. Levy
Jarnac Observatory, Inc.
Vail, AZ, USA

ISSN 0067-0057 ISSN 2214-7985 (electronic)
Astrophysics and Space Science Library
ISBN 978-3-319-19877-4 ISBN 978-3-319-19878-1 (eBook)
DOI 10.1007/978-3-319-19878-1

Library of Congress Control Number: 2015950465

Springer Cham Heidelberg New York Dordrecht London

Printed on acid-free paper

Springer International Publishing AG Switzerland is part of Springer Science+Business Media (www.springer.com)

For Wendee. May you always reach for the stars with me.

Preface

This investigation is a two-part study of allusions to events in the sky in English writing from 1572, the year of Tycho Brahe's great supernova, to a time shortly after the first use of the telescope in astronomy in 1610, and during the Victorian period featuring Tennyson and Hopkins. The early modern period is a time during which specific phenomena of the night sky increasingly appear in early modern English Literature, and the mid-nineteenth century era of Tennyson and Hopkins is a time featuring seismic changes in humanity's understanding of the Universe.

Although much has been written about the changing cosmic philosophies of the Renaissance, I will explore a different line of inquiry—a selection of actual events in the sky as they appear in the literature of that time and of the mid-Victorian period. This text emphasizes a selection of events, like new stars or supernovae, comets, meteors, and eclipses, which took place between the autumn of 1572, when the first "blazing starre" in over 500 years thrilled viewers of the night sky, and Galileo's discoveries with his telescope in 1610; and later celestial events. I chose these periods because they offered an unusually large number of such events, specifically two supernovae within our home galaxy and 20 comets, whereas during the last 30 years of our own time—a more typical period—there were seven bright comets and no supernovae within our galaxy. These unusual events are referred to amid a rich background of allusions to more common events like sunrises, sunsets, and meteors.

I will approach this topic by selecting passages in works of literature that might correspond to specific events. This study will take advantage of recent technological advances that have given us a much clearer understanding of what events actually took place in the sky and how they were exploited by contemporary authors. Computer programs have recently become quite sophisticated in taking us back into time to show us the positions and magnitudes (brightnesses) of supernovae, the paths of comets as they crossed the sky, and where, when, and to what extent eclipses were visible over specific cities. I use this information in my discussion of well-known references to eclipses such as Shakespeare's *King Lear*, but also in little-known writings like Thomas Nashe's *A Wonderful, strange, and miraculous, Astrological Prognostication for this year of our Lord God, 1591*, in which an eclipse of the Moon that December is described. As a specific study in English literature, this research might appear to begin as an exercise in annotation; i.e. a series of explanations of the astronomical references that appear in the literature of these two periods. But it offers much more: When seen in the context of literature *and* the night sky, the connections between literature and the sky that it reveals will open the offerings of the night sky of those eras to the people of our own time.

The relationship between literature and science is an increasingly robust field of enquiry. As part of my research for this dissertation, in the spring of 2004 I offered a course in astronomy and literature at Arizona State University where, for one semester, I worked closely with a group of undergraduate students studying ideas in this field. The course was called "From Shakespeare to Space" and took a broad look at how different writers used the ideas of astronomy. For example, did Shakespeare have specific celestial events in mind when he wrote *King Lear*? Was Van Gogh's *Starry Night* inspired by a scientific observation of a distant galaxy by the Third Earl of Rosse using his giant reflector, then the largest telescope in the world? This course explored the union of two of the most basic areas of study that a university can offer in the arts and the sciences through a discussion of specific instances in astronomy and literature. The course also included seminar presentations by each student. I was impressed at how adept these students were in linking their own studies to discrete arenas in English Literature. By bringing together the arts and sciences, this course tried to inspire its students to see a vast picture in planning their future, whether that future is to be in arts, science, or in teaching youngsters to live creatively in our complex world.

The course I gave helped to inspire my students to probe more deeply into the broader interpretations that are possible when English Literature is related to some other field, particularly a scientific one. Astronomy is a broad field, but the night sky subset is narrow enough to provide a benchmark to focus these interpretations. At the undergraduate level, I encouraged my students to let their minds wander, focusing as far afield as the Victorian and modern periods. The frequent references to the Moon in Charlotte Bronte's novel *Jane Eyre*, as well as similar references in Tolkien's *The Two Towers*, provided a good sense of thematically relevant coloring to those narratives. Most students chose a specific period of literature on which to prepare a seminar, although some students went even farther afield, relating the literature to painting or music.

This course was complemented by a public lecture I gave as part of my role as John Rhodes Chair in Public Policy and American Institutions at ASU. My topic, "Space Policy and America's Future: Igniting the Imagination" was designed to show how literature relates to scientific endeavors like the night sky, and even to space travel. The lecture was structured to show that, as the United States pondered its future in space after the loss of the space shuttle Columbia in 2003, it would be advantageous to take a broad view of what typical Americans are interested in. The result of that process was the beginning of the program to retire the space shuttle completely, then replace it with fresh vehicles and rockets designed to return humans to the Moon and to send them further into space, perhaps to Mars, or to a nearby asteroid that, though it orbits the Sun, could someday collide with Earth.

The dream of going to visit other worlds lies deep in human literary history. It takes us back in time to the ideas expressed by writers and scientists dating at least as far back as the Roman author Lucretius, whose *De Rerum Natura* was a guide to how the natural world works, and an anthem on why it is important to explore that world. In the world of politics, that means a physical exploration, using vehicles and other technology to whisk humans from the Earth to the Moon and Mars. But there are other ways to conduct that explorative journey.

In 1970, as an undergraduate at Acadia University, I was inspired by Roger Lewis to seek out a connection between the night sky and poetry through an essay "Elements of Science in Tennyson's In Memoriam." Years later, Norman MacKenzie, at Queens University, directed me to Gerard Manley Hopkins's small poem about a comet that formed the basis of my Master's thesis "The Starlight Night: Hopkins and Astronomy," a project that now forms several chapters of this book, which has expanded those earlier studies to include a broader array of English authors from the pre-modern era. While *In Memoriam* directs us specifically to the discovery of Neptune, and to the then-popular nebular hypothesis of the formation of our solar system, generally the allusions in the early modern period are less specific.

I began this journey in a spirit of learning and fun, with the obvious references to "these late eclipses" of the Moon and Sun in 1605. I quickly found a confirming letter from King James to his close friend and advisor, Sir Robert Cecil, in which he playfully discussed the effects of the eclipse as seen in England. King James had a serious policy against the practice of astrology. The Stuart King also presided over two of the most important books ever published in the English language, the translation of the Bible into English in 1611, and the First Folio edition of Shakespeare's works in 1623. In a sense, these two books testified to the emergence of England's leadership in the arts and in religion that helped to cement the nation as a global power.

I shared these ideas as a guest lecturer in astronomy and literature at other universities, notably Southern Illinois University at Carbondale, Mount Allison in Canada, and at a return visit to Acadia University. Those three institutions afforded me the opportunity to share my passion for literature and the night sky with students who were just starting their own career paths. Those lectures differed in some fundamental respects. At SIUC, I interacted with students throughout their honors program, tried to connect literature and the night sky in the early modern period to the areas in which they were interested. At Mount Allison and at Acadia, where the students were specifically studying science and literature, I brought my perspective to them. In this context, I admit that this perspective was more personal than scholarly, and it did trace my early attempts at a union of literature and astronomy.

Vail, AZ, USA David H. Levy

Acknowledgements

In November 2013, while attending an astronomy conference in Tucson, I stopped by the Springer exhibit and met Nora Rawn, one of the editors at Springer. We started talking, about nothing in particular at first, then I mentioned that Springer had published *The Sky in Early Modern English Literature*, based on the Ph.D. I had completed in 2010 at the Hebrew University. Before I knew what I was saying I opened my big mouth and suggested a second edition, and before I left the convention center I decided that somehow I was going to add the MA thesis on the same topic but featuring the Victorian poet Gerard Manley Hopkins. I knew later that day that I'd have to come up with some writing that connects the two far-flung eras of literature. And before I fell asleep that night, I decided to add also a chapter on Tennyson. I loved doing the thesis, which took as long to propose as it did to write. I loved turning the thesis into a book, and I've loved every minute of preparing this new edition. Perhaps the idea to produce this book evolved from an essay I wrote for a University of Tampa project about Literature and Science, edited by Dr. Judy Hayden. During that process several improvements were made that increased my motivation to pursue a new edition. As this new book comes to fruition, I doubt I'll ever fall out of love with its subject.

The idea of relating astronomy to literature has been in my mind for almost as long as my interest in astronomy. I owe that happenstance first to my brother Richard, with whom I lost an argument back in 1957 when I was 9 years old. My response was to pick up a copy of the nearest book I could find, which happened to be the beautiful blue-covered Yale Shakespeare copy of *Hamlet*, and throw it at him. If my father hadn't walked in at just that moment, the incident would likely have ended there. Dad calmly picked up the small book from the floor, handed it to me, and said, "David, I know you're going to have arguments with your brother. You'll even throw things at him sometimes. You can even throw a stone at him. But don't ever throw a book at him or anyone."

Smarting from this putdown, I challenged my father. "Why not?" I will never forget his carefully phrased answer: "Because books are friends. It is as if this author invited you into his living room, ushering you into a far-off time and place; as if the author tells you that through his book you will learn about what it was like to be alive in his time." When I began this project I stepped back into time, under the sky of early modern England, just as Dad had suggested so long ago.

After a long period of doing nothing about this curiosity, I found it resurging to life on the night of April 23, 1976. With a group of friends at the Montreal Centre's observatory of the Royal Astronomical Society of Canada, I was observing the maximum of the Lyrid Meteor shower during Session 2887M (each one recorded in sequence in my observing logs). I saw perhaps 15 meteors that night, and I got the idea that, as I prepared to enter graduate school at Queen's University, I might want to write a thesis about how one of these poets read the night sky.

The next day, I met Dr. Norman MacKenzie at his English Department office in Watson Hall at Queen's University. On that spring Saturday in Kingston, Ontario, as we prepared to discuss my future in his department, I mentioned my "newfound" interest. Immediately he uttered a single word, "Hopkins." He explained how his favorite poet had a passion for the sky, and then he called to my attention one of his early poems, a fragment called "I am like a slip of comet." I devoured that poem like a hungry animal. I wasn't sure at that early time how I could make it work, but I loved the poem. Years after I had completed my M.A., he said in a lecture that he never witnessed anyone get as excited about a work of literature as I did about the Hopkins comet poem.

All these decades later, Hopkins's comet poem still calls to me just the way it did in the spring of 1976. I enjoy reading it over and over, and have developed a way to read it while watching a film showing Comet Hyakutake, one of the best comets of the last century, as it crosses across the northern sky.

Comets, and Hopkins's poem about one particular comet, evolved into a successful MA thesis at Queen's. After that I decided to delay proceeding to my Ph.D. About 5 years after graduation the idea resurfaced and I quickly buried it. It arrived while I was driving Brian Marsden, the longtime director of the International Astronomical Union's Central Bureau for Astronomical Telegrams, home from an evening at Kitt Peak National Observatory. Five years later the idea surfaced again and I quickly buried it. When it surfaced a third time I was probably observing with the Shoemakers at Palomar Observatory. At two additional 5-year intervals the idea returned, and each time I rejected it. The fifth time this happened I was freshly married to Wendee. I mentioned it to her and quickly added that I would bury it as always. "Before you reject it this time," Wendee responded, "let's discuss this. Maybe you keep thinking about it because you want to finish what you began so long ago." That conversation matured into letters, into my choice to write a dissertation on the night sky in the early modern period, and into my selection of Dr. Lawrence Besserman as my thesis director at the Hebrew University of Jerusalem. Dr. Besserman was extraordinarily helpful throughout the long and challenging proposal process and even more so during its research and writing phases. Working with him was a sheer pleasure. During the difficult proposal process, my committee soundly killed the plan, but when I suggested that we form an international committee consisting of scholars I respected throughout the world, the Hebrew University approved enthusiastically. These people have taken time off their own busy lives to help move my project along. Specifically, Frederick Williams of Southern Illinois University of Carbondale did a superb job as associate dissertation director. I also wish to thank Janine Rogers of Mount Allison University, Karen Bamford and Robert Lapp also of Mount Allison, Larry Lebofsky of the University of Arizona, Martin Rice of the University of Pittsburgh at Johnstown, Eli Maor of Loyola University in Chicago, David Mowry of the State University College at Plattsburgh, Allan Chapman of Wadham College, Oxford University, David DeVorkin of the National Air and Space Museum, and Ilan Manulis from Israel's Weizmann Institute of Science, all of whom provided copious amounts of valuable help and advice. And they all encouraged me to keep going even as I slowly recovered from a stroke early in 2007.

For my earlier work on Hopkins, specially revised for this new edition, Dr. Norman MacKenzie of Queen's University acted splendidly as supervisor. His patience and understanding were as invaluable to me as the depth of his knowledge on both Hopkins and the astronomy of his age helped transform that part of the project into a worthwhile addition to this edition. His passing in 2004 remains a deep loss. It was a special honor to have Dr. A. Vibert Douglas, one of Canada's most famous astronomers, as a reader for the Hopkins section. Her suggestions went far beyond her formal interest in astronomy, and the lively interest and care that she projected helped make it an experience to be treasured. Dr. Joseph Ashbrook of *Sky & Telescope* magazine prepared an ephemeris of the nonperiodic Comet Tempel (C/1861 N4 according to the new reckoning begun in 1995) which added important evidence that it did play a role in Hopkins's comet poetic fragment. Michel and Lorraine Payette helped with the calculations for the planetary positions that appear frequently in the Hopkins section. The staffs of the Libraries at both Queen's, where I completed the Hopkins M.A., and the Hebrew University where I completed my Ph.D., both provided first class assistance in the special needs both these projects demanded. An unusually obliging librarian at the University of British Columbia confirmed a comet reference while I waited patiently over the telephone during an expensive person-to-person long distance call in the autumn of 1979.

Three people deserve special thanks. Eli Maor expertly translated the opening pages of the thesis into Hebrew, and Roger Lewis, now professor emeritus at Acadia University's Department of English, gave detailed help and guidance at every stage and was particularly helpful with the chapter on Tennyson, which began as an essay in his Victorian poetry class at Acadia in late 1969 and 1970. (It was wonderful to observe how the original essay, which he graded as an A minus, surged to an A when he graded the most recent version. He also added that with my additional experience over the years, it was to be expected that the grade would increase.) Finally, this project would never have been started, let alone completed, had it not been for my wife Wendee's constant encouragement throughout the years and active assistance in formatting and in catching typing errors and inconsistencies. Without her critical eye, the book simply could not have been completed.

Contents

Introduction

Imagine a mostly cloudy October morning in southern England. In busy London, people were going about their daily business, and far out of town, James I, King of England, was probably enjoying a hunt (Jeayes 161–62, Harrison 235). A slight lessening of sunlight began in the late morning, but it did not attract much notice until noon when the sky began to darken noticeably and rapidly, much more deeply and rapidly than it would have from the onset of clouds. Between 12:40 and 1:00 p.m. the land was bathed in a twilight hue. Through breaks in the clouds, the Sun peeked through not as a bright yellow ball, but as a thin curved line of light. The date, according to the Julian calendar then used in England, was 2 October 1605 (12 October by the then-new Gregorian Calendar, not yet in use in England), and southern England was experiencing a solar eclipse. As the Moon continued moving eastward across the Sun, it lessened and then abandoned its apparent grip on the Sun just after 2:00 p.m. (Espenak, Pingre). Many Londoners were not surprised at the occurrence of this near-total solar eclipse over their city; they might have read of it in Dade's Almanac (n.p.). It is also possible that some Londoners realized that the event was coming after watching a performance of *King Lear*, complete with a remark about "these late eclipses in the sun and moon," earlier that year in the Globe Theatre (Schoenbaum 253).

The October eclipse was the last of a series of three eclipses, two lunar and one solar, to occur over London in 1605, and it offers a focus point for this investigation of convergence between the two fields of endeavor of English literature and the study of the night sky. That eclipse also provides an impetus to follow Recorde's advice from 1552 to understand the sky: If Reasons reach transcend the Skie, Why should it then to earth be bound? The wit is wronged and led awrie, If mind be married to the ground. (Recorde, 3)

This eclipse represents one of those rare times in history when literature, culture, and science come together. It allows a reconstruction of a few hours of time almost 400 years ago with the help of a variety of scientific and literary sources. The solar eclipse did happen at the time specified. Whether *King Lear*, with its ominous debate about the predictive value of eclipses, had already been completed and performed in London, or whether Shakespeare was completing its composition at the time, is less certain.

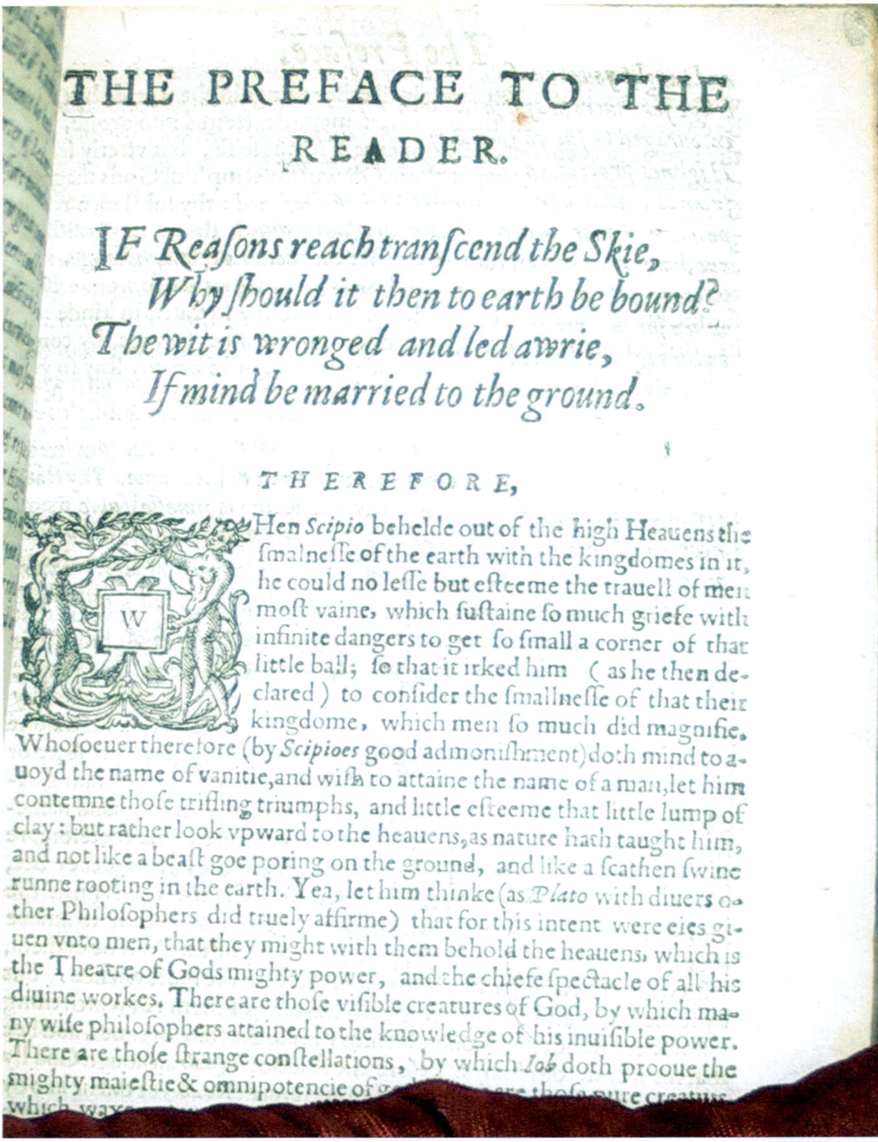
THE PREFACE TO THE READER.

IF Reaſons reach tranſcend the Skie,
Why ſhould it then to earth be bound?
The wit is wronged and led awrie,
If mind be married to the ground.

THEREFORE,

WHen *Scipio* beheld out of the high Heauens the ſmalneſſe of the earth with the kingdomes in it, he could no leſſe but eſteeme the trauell of men moſt vaine, which ſuſtaine ſo much griefe with infinite dangers to get ſo ſmall a corner of that little ball; ſo that it irked him (as he then declared) to conſider the ſmallneſſe of that their kingdome, which men ſo much did magnifie. Whoſoeuer therefore (by *Scipioes* good admoniſhment) doth mind to auoyd the name of vanitie, and wiſh to attaine the name of a man, let him contemne thoſe trifling triumphs, and little eſteeme that little lump of clay: but rather look vpward to the heauens, as nature hath taught him, and not like a beaſt goe poring on the ground, and like a ſcathen ſwine runne rooting in the earth. Yea, let him thinke (as *Plato* with diuers other Philoſophers did truely affirme) that for this intent were eies giuen vnto men, that they might with them behold the heauens, which is the Theatre of Gods mighty power, and the chiefe ſpectacle of all his diuine workes. There are thoſe viſible creatures of God, by which many wiſe philoſophers attained to the knowledge of his inuiſible power. There are thoſe ſtrange conſtellations, by which *Iob* doth prooue the mighty maieſtie & omnipotencie of g[illegible] are thoſe pure creatures which wax[illegible]

Fig. 1 This photograph shows Robert Recorde's advice that the value of reason is so great that it can transcend the sky. Photograph by the author

Literature and events in the sky are subjects not normally studied together in university programs, but their occasional linking together offers a unique investigative opportunity. We can begin to imagine the person watching *King Lear* suddenly being drawn by its references to celestial events to talk of eclipses and memories of new stars he or she has seen in the sky.

In examining the extent to which the phenomena of the night sky between 1572 and 1620 appear in early modern English Literature, I am guided by an axiom that events in the sky, like comets, novae, and eclipses, can help us understand the literature of the time, and that conversely, the literature of the time can aid in our understanding the events of the sky. Literature and the sky complement each other. This investigation is not, however, specifically about the changing cosmic philosophies of the time, a subject that has been covered well by critics since Marjorie Hope Nicolson and Cumberland Clark.

From an interest based on astronomical observation more than philosophy, I will explore a different line of inquiry—a selection of actual events in the sky as they appear in the literature of the time, with a view to studying them in order to shed light on both fields. The events chosen took place between the autumn of 1572, when the first "blazing starre" in over 500 years exploded into the night sky, and the early telescopic discoveries in 1610.

I chose this period because it offered an unusually large number of such events, specifically two supernovae, twenty comets, and several well-observed eclipses, whereas during the last 30 years of our own time—a more typical period—there were only seven comets and no galactic supernovae. These unusual events are recognized by English writers of the early modern period amid a rich background of allusions to common events like dawn, meteors, and the general appearance of the night sky. Frequent notice of celestial events by pre-modern English writers resonates with my own long career as a "night watchman," a searcher of the night sky.

Opening Thoughts

Four hundred years after the Shakespeare family celebrated William's first birthday in England, on September 18, 1965, Kaoru Ikeya was peering through the eyepiece of his homemade 8-inch diameter reflector. A worker in a piano factory in Japan, he was enjoying some free time with his telescope, searching for comets. He had already discovered two, his first in January 1963 and his second in the late summer of 1964. On this late summer morning of September 18, 1965, he spotted a little spot of haze in its field of view. Ikeya knew the sky, and he was fairly certain that the object was an anomaly, that it was not permanently in that position. His star atlas showed nothing, and a look through the eyepiece a short while later convinced him that he really was looking at a comet, for the fuzzy patch of light had moved a barely detectable distance in the time since he had first spotted it. Ikeya correctly identified this new object as a comet, and lost little time in sending a telegram to Japan's Tokyo Observatory. Just 1 hour later, Tsutomu Seki, a guitar instructor at the time, found the same object.

Comets were arguably as important to Shakespeare four centuries ago as they were in 1965 to amateur astronomers like Ikeya and Seki. We know this because of the number of times he mentioned comets; for example, during the summer of 1596, Shakespeare might have observed a bright comet himself. First seen on July 11, this comet was as bright as some of the brightest stars in the sky. In any event, he composed these words to be spoken by Caesar's wife Calpurnia in *Julius Caesar*:

When beggars die, there are no comets seen
The heavens themselves blaze forth the death of princes.
(*Julius Caesar*.2.2.30–31).

One of Shakespeare's most famous references to comets, this particular couplet is historically significant because during the Octavian games in the summer of 44 BC, held in memory of the assassinated Caesar, a bright comet did indeed track through the northern sky and was widely believed at the time to represent the soul of the assassinated Caesar on its way to heaven. The comets of Calpurnia's dream were premonitions that appeared before her husband's assassination, but the historical comet was reported to have appeared (although there is some doubt that it appeared at all) 2 months *afterwards*, with a tail perhaps half the length of the Big Dipper, as the comet crept through the northern sky. The opening three lines of *1 Henry VI* summon his audiences to remember the great comets that have visited the region of space, including one that rounded the Sun in April, 1590. Discovered by Tycho Brahe, that particular comet became as bright as the star Capella (G. Kronk, 9). Alternatively, there might have been an outburst of Comet 17P/Holmes on February 10, 1591, a week before its perihelion passage (Z. Sekanina 4–17). The comet was definitely there, but it is not known whether it was bright enough to be visible. This particular comet has undergone at least two major outbursts in historic times, one around the time of its discovery by Holmes in 1892, and again in 2007. Either way, the three opening lines of *I Henry VI*, dated 1591, expressed the idea of comets as omens:

Hung be the heavens with black! Yield, day, to night!
Comets, importing change of times and states,
Brandish your crystal tresses in the sky…(*I Henry 6*.1.1.1–3)

(The Oxford editors suggest the possibility that Shakespeare collaborated with other writers on this play, and that the opening might have been composed by Nashe. I do not subscribe to this interpretation, but even if Nashe did write the opening lines of *1 Henry VI* the comet allusion remains valid, since Nashe also had an interest in comets.)

Neither Shakespeare nor Nashe was the first writer to express an interest in the night sky; this tradition goes back much farther than Dante, whose *La Vita Nuova* (*The New Life*) explores how

The sun ceased, and the stars began to gather,
And birds dropped in mid-flight out of the sky;
And earth shook suddenly; …

One of the earliest known writers, Enheduanna, who lived 4200 years ago (2285–2250 BCE) as the daughter of King Sargon of Akkad, alluded to the sky frequently in her writings:

> Inside the light is dim
> Even moonlight (Nanna's light) does not enter
> (*Temple Hymn 7*, 10–11).

In her *Temple Hymn 42*, Enheduanna metaphorically speaks to the temple itself as she summons

> This shining house of stars bright with lapis stones
> Has opened itself to all lands (1–2).

In her *Exaltation of Inana*, Enheduanna designates Inana as "powerful one of heaven and earth, you are their Inana…" (11–12).

What comets did Enheduanna witness during her lifetime? If our understanding of its orbit is just slightly off, it is not outside the realm of possibility that the Akkadian princess observed Comet Hale-Bopp at its last return before its modern visit in 1997. No doubt, she did see eclipses of the Moon and the Sun, and possibly at least one bright comet.

As long ago as Enheduanna lived and composed, the wisdom of her writing probably was not a factor to the two young cometeers from Japan who discovered the comet of 1965. At discovery, its brightness was about six times fainter than the faintest star one can normally see without a telescope. Its motion gave the first clue of what was to come; it was moving almost directly toward the Sun. Within a few weeks, this visitor from space rounded the Sun, become brighter than the full Moon, displayed a beautiful tail 70 million miles long, and started to move back away toward the dark void of interplanetary space from where it came. It was probably the brightest comet of the twentieth century.

In the sense that the Great Comet of 1965 and the several bright comets of Shakespeare's time all passed close to the Sun, these mighty comets were similar. Ikeya's and Seki's prompt reporting of it continued a tradition that has persisted as far back as Shakespeare's time during which it was still not completely known whether comets flew within our solar system or were merely appearances in our atmosphere. Tycho Brahe, who independently discovered the comet of 1577 and who probably did discover the comets of 1582 and 1590 (Kronk, 9), concluded from observations sent to him from observers across Europe that these comets were farther from Earth than the Moon.

Over the centuries, astronomy has evolved to the point that professional astronomers and many amateur astronomers as well have so narrowed their own fields of inquiry that they no longer directly look at the sky at all. In the sixteenth century, an era without electricity and light pollution, television, and computers, skywatching was more common and democratic. Educated people were expected to learn something of the night sky, and Bacon's *Advancement of Learning* (125, 141, 158) was encouraging them to do so. Writers might include routine sky events, like a meteor or a bright planet in the evening sky, even a sunset, in their writings, and they could expect their readers to enjoy and understand these references.

This topic is approached through select passages in works of literature that allude or refer to specific events. The study takes advantage of recent technological advances that have given a much clearer understanding of what events actually took place in the sky and how they were exploited by contemporary authors. Computer programs have recently become quite sophisticated in taking us back in time to show us the positions and magnitudes of supernovae, the paths of comets as they crossed the sky, and where, when, and to what extent eclipses were visible over specific cities. This information will be used to discuss well-known references to eclipses as well as those in lesser-known writings. For example, the "late eclipses in the sun and moon" in Shakespeare's *King Lear* should be better known to most high school students than Thomas Nashe's *A Wonderful, strange, and miraculous, Astrological Prognostication for this year of our Lord God, 1591*, in which a rare total eclipse of the Moon that December is described. That particular event turns out to involve a rare coincidence; in addition to entering the shadow of the Earth, the Moon also passed in front of the planet Saturn, resulting in an occultation, at the same time.

This book offers more than a series of explanations and interpretations of the astronomical references that appear in the literature. When seen in the context of literature *and* astronomy, the connections between literature and the sky that it reveals will open the night sky of that period to the people of our own time, and broaden our understanding of the literary works studied as well.

Examples of the primary sources used in Part I focus mainly on Shakespeare, whose celestial references surpass all the other authors put together; in fact by one count there were at least 205 separate allusions to the sky, or to a celestial object, in his canon. Nowhere do we find a statement by Shakespeare that he enjoyed observing the night sky, but his works undeniably testify to a passionate interest. For example, *King Lear's* discussion of eclipses could refer to the extraordinary pair of eclipses that occurred in the early autumn of 1605. Eclipses are referred to explicitly in the play (*KL* 1.2.99–127), and although some critics propose that he had no particular eclipses in mind, I will provide evidence that he did mean the two in 1605.

With these events as a basis, Shakespeare proposes the idea that humanity has a cosmic relation. Part of that relation involves the public debate between astronomy, the study of the stars, and astrology, which was defined as the use of the stars to predict human events. Since Shakespeare intended his plays, and the ideas contained within them, to be presented before wide and diverse audiences, he not only spoke to his time but also helped to shape it. Shakespeare's plays were to his time as the most popular television shows are to ours. Thus, in *King Lear* Shakespeare took advantage of his viewers' familiarity with the eclipses and two new stars to debate their possible effects on humanity and the affairs of state.

This book aims to uncover a cultural interest in the night sky that was integral to the culture of the day in early modern English literature. To accomplish this I explore passages involving the night sky by both canonical and obscure writers, with an aim to recontextualizing those passages as they relate to astronomical events and objects. Ultimately, these passages illustrate how the literature of the time acts as a mirror that reflects the interest of the people of that time in the sky and its special events; also they will help show how an understanding of cosmic events can lead to a better appreciation of the literature that uses them.

This is a study of historical genesis that fits into the two disciplines of literature and astronomy in several ways. One involves using the literature to help us see the sky through the eyes of the people who lived under it. Another involves dating. In writing about the influence of external factors upon the creation of literature, much has been noted about underlying political and social conditions. These circumstances can be used to help date plays in the same sense that the Porter's speech in *Macbeth* (2.3.1–20) alludes to the trial of Guy Fawkes in 1606 and therefore can better determine the play's date of composition. However, thus far much less attention has been paid to the influence of natural events that can be pegged to a specific time just as accurately, like the two eclipses in the fall of 1605 that help date *King Lear*.

Beyond dating, events in the sky can help us interpret specific passages of literature and help us to a clearer understanding of what was in the creator's mind. As an example, Shakespeare's vivid description at the opening of *Hamlet* of a "star that's westward from the pole …" (1.1.35–38) is consistent with the appearance of the very bright *stella nova* that appeared in November 1572. In this sense, this research will enable students of the period to see the literature in the context of actual events in the night sky.

In the world of astronomy, English literature can function as a window that allows those of us familiar with the sky of our own time to visit and understand the sky of an earlier time. While it is possible to build a comprehension of the details of an event by looking it up in a contemporary almanac, one gains a coherent sense of how these events were perceived by the general population through its literature. The eclipse scene in *King Lear* (1.2.99–127), together with King James's Letter to Robert Cecil about the 1605 solar eclipse (Akrigg 264–66), provides such insights.

Using this approach, I discuss various works of literature in which references to the sky appear. In some cases (i.e. the lunar eclipse of 1591), I show how a particular reference sheds light on how a particular aspect of the sky was perceived; in others (*King Lear's* eclipses), I show how events in the sky are a major factor in character and plot development. In both senses, this adds to our understanding of the night sky, through the literature that references it, at a time when an impressive series of comets, new stars, and eclipses were drawing attention to its physical nature.

Structure

The astronomical timeline in the period 1572–1620 is the foundation for Part I. The chapters are organized according to type of event (stellar explosions, comets and meteors, eclipses, conjunctions, and the telescope) and within each chapter the references generally begin with Shakespeare, whose work consistently points to nature in general and the night sky in particular, and then proceeds to the works of his contemporaries.

(A) General Description. This section outlines the idea of how two subjects—English literature and the night sky—can be merged into a single thesis to produce results that would be useful to other researchers.

(B) Approach. I approach this topic by selecting passages in works of literature that respond to specific events in the sky. My primary sources include such astronomical works as Abraham Fleming's *Prognostication of Blazing Starres*, a creative translation of a Pontanus poem that emphasizes the differences between stars that "offer not to change" their positions in the sky (novae) and those that "take their course unto the east" (comets) (Nausea A.v.). I will dem-

> onstrate that Fleming's work appears to follow directly a series of comets that appeared in the sky between 1577 and 1602. Other sources include Tycho Brahe's treatise on the Supernova of 1572, Shakelton's treatise about the comet of 1577 and its significance, Nausea's discourse on comets, and Bainbridge's thesis on comets. These sources are wonderful examples of writers who have tried, in ages past, to inspire their readers to dig more deeply into the unusual events occurring at the time, they have rarely been cited in earlier documents on science and literature, but they were important contributors to this literary culture of thinking about the sky.

My literary sources begin with Shakespeare as the keystone author. *King Lear*, to cite an important example, contains passages that appear to respond to a series of three specific eclipses in 1605 (*KL* 1.2.99–127). The series included a near-total eclipse of the Moon (more than 99 % of the Moon was covered in the Earth's shadow) in the after-dinner hours of 24 March (O.S.), a partial eclipse before dawn on 27 September, and a partial eclipse of the Sun only 2 weeks later on 12 October during which most of the Sun was covered by the Moon as seen from London (Espenak, Eclipse Home Page). I will provide fresh evidence that these eclipses are referred to explicitly in the play (99–127). In addressing these events, Shakespeare helped open the minds of his audience to new ideas about humanity's relation to the cosmos. As a shareholder in his company, Shakespeare most likely intended his plays, and the ideas contained within them, to be presented before wide and diverse audiences; he not only spoke to his time but also helped to shape it. Thus in *King Lear* Shakespeare took advantage of his viewers' puzzlement over the eclipses to consider that *Lear*, as John Danby suggested years ago, is a drama of ideas, a play specifically dramatizing possible effects on humanity and the affairs of state.

In *King Lear* the meaning of the word "Nature" evolves to a Baconian *Novum Organum* of Elizabethan thought brought to life through drama (Danby 15). Danby is outdated by today's standards, but some of his ideas have become current again. Edmund is animated by the idea of nature as he announces

> Thou, Nature, art my goddess; to thy Law
> My services are bound (1.2.1–2).

In this early moment of the play, notes Dan Brayton more recently, "what Lear cannot see, and what is perhaps glimpsed by Cordelia and Kent in their reactions to Lear's living will, is that the process of division initiated with the display of the map will become uncontrollable, as the play proceeds to leave nearly all of its major characters propertyless, bereft, or dead" (Brayton, 402–403).

Soon after, the eclipses are invoked, and in a speech filled with concern for the future, the Earl of Gloucester argues: "These late eclipses in the Sun and Moon portend no good to us. Though the wisdom of nature can reason it thus and thus, yet nature finds itself scourged by the sequent effects. Love cools, friendship falls off, brothers divide" (1.2.101). As soon as Gloucester exits, Edmund mocks his father's reasoning: "This is the excellent foppery of the world, that when we are sick in fortune, often the surfeits of our own behavior, we make guilty of our disasters the sun, the moon, and stars, as if we were villains on necessity" (1.2.115–127).

Edmund's speech, with its skeptical-rationalist perspective, introduces a central issue in *King Lear*, a theme that reached an audience that had the eclipses in mind. "Nature" in this play is painted with a broad and all-inclusive brush, including events in the natural world as well as actual natural events happening in the world of Shakespeare's time. Danby's great contribution was to consider Shakespeare's interest in Nature in light of the words spoken by characters such as Edmund. I believe that Danby's theme is as relevant in today's world as it was when he first proposed this idea in 1948. The fact that we now know more about Shakespeare's interest in natural phenomena than we did then serves to support my point. Even the names of *King Lear*'s major characters—Albany, Gloucester, Cornwall, France, Burgundy, and Kent are all named for specific geographical places—display a sharp Shakespearean focus on the physical map of Europe, not unlike the map that Lear presents ("Give me the map there") at the play's opening (*KL*.1.1.36).

The eclipses were not the only extraordinary celestial events taking place in 1605. A "blazing starre" or *stella nova* that appeared in the constellation of Ophiuchus in 1604 was still bright in 1605 and was referred to specifically by Jonson in a passage outlining a series of unusual (and real) contemporary events used as omens in *Volpone* (2.1.47–50).

We now know that Kepler's star, as it is now called, was a supernova, the result of a star whose central core collapsed, sending the rest of its matter hurtling into space and shining as brightly as the combined light of all the other stars in our galaxy combined. Although no one was aware of the true nature of the star at the time, it was easy to be aware of its presence as it shone brightly enough to be visible "like a burning light" (Christianson, 275). It is a further contention of this dissertation that the two fields relating to the stars, what we now refer to as astronomy and astrology, became recognized as different fields during Shakespeare's time, partly because of the unusual celestial events going on. There were two ways of studying and interpreting these events, that of *astrologia naturalis* and the other of *astrologia judicialis*. The former holds that heavenly bodies have influence on physical manifestations of our lives, like weather and physical matter. The latter suggests that these bodies influence human destiny (Sondheim, 243–259).

The Major Celestial Events of the Period

That this period is a rich one for the sky is evidenced by the following parade of celestial events:

1. The supernova of 1572: We now know that Tycho's star, as it is called, was the first supernova visible from the Earth in almost 400 years, the result of a star whose central core collapsed, sending the rest of its matter hurtling into space and shining as brightly as the combined light of 200 billion suns.
2. Tycho's comet of 1577: Discovered by Tycho Brahe (among others) on 13 November 1577, this comet was brighter than any seen in the previous century.

The comet was followed by a procession of 20 other comets, all visible to the unaided eye in this pre-telescopic era.

3. The Great Conjunction of 1583: This event occurring once every 20 years is a closing together of Jupiter and Saturn.
4. The Lunar Eclipse of December 1591: While the eastward-moving Moon was still in partial eclipse, it passed in front of Saturn, an event which has not occurred since then.
5. Eclipses, particularly in 1598 and 1605: While clouded out in London, the effects of the darkening of the sky during the 1598 eclipse were recorded by Queen Elizabeth I's aide John Dee. More than 99 % of the Sun was covered at maximum eclipse; it was even closer to total in Stratford, and was total 160 miles away near Neath, England. From London, Dee wrote: "Feb. 25th, the eclips. A clowdy day, but great darkness abowt 9½ mane" (Dee 61). There were actually three eclipses in 1605: a total lunar eclipse on the evening of April 3, a partial lunar eclipse on September 27, and a near-total solar eclipse on October 12.
6. A heavy shower of meteors in November 1602: The Chinese text *ThienWen-Chih* records that on November 6, "Hundreds of large and small stars flew, crossing each other." The Korean text *Munhon-Piko* records that on November 11, "Many stars flew in all directions." (http://www.amsmeteors.org/comets/meteors.showers/leonidancient.htm, Levy, 12) Although this was a significant event in some parts of the world, it was not visible to as great an extent over England. It is possible that the shower was also seen from England, in much the same way as the 1998 Leonid meteor shower was intense, for more than 36 hours, all over the world. (It is also obvious that one of these two references has an incorrect date.)
7. A second supernova in 1604: This *stella nova* in Ophiuchus outshone every other star in the sky and was brighter even than the planet Jupiter. It remained bright throughout 1605. The appearances of two observable supernovae in our own galaxy within a single human lifetime are unprecedented; no supernova event visible to the naked eye has appeared since.
8. The Telescope, 1608: Although there is controversy over who actually was the first to point a telescope to the sky, no doubt exists that Galileo was the first to observe the Moon, Jupiter, Venus, and the Sun, to record carefully these observations, and then to publish them widely. Galileo made the sky more democratic; anyone with a telescope could see what he saw. But he also started astronomy's road to where it is now, where a class of professional astronomers no longer "looks" at the sky but digests it in the form of computerized data.

The Emerging Role of Compilatio

In her article "Lover, Poet, or Astronomer: Collecting Stars and Poems with David H. Levy," Janine Rogers of Mount Allison University posits that literature can influence the way the night sky is read. Building on philosophers like Plotinus and

Marsilio Ficino, who wrote that "The celestial configurations are like the letters in a book which explain the divine concepts" (Garin, 67) she expanded on the idea of "reading" the night sky as a book, for it provides a rallying point for the ideas of observation, and supports my rationale for interpreting appropriate fragments of literature that help us "read" a particular aspect of the contemporary night sky.

Rogers introduces the term *compilatio*; a relative of *complication* which represents "the action of folding together" (OED), this refers to bringing together two disparate fields of study that would, on each surface, appear to have little in common. *Compilatio* is seen as a way of "building a greater understanding of the world through layering of several texts together." As a specific medieval concept, the term invites a collection of material from different sources, and then reassembles them in a new and different framework. (Rogers, personal communication, January 2008, 13 July 2008, 16 July 2008)

Benefits from this research work both ways; sky references reveal new meanings in the literature, and they also help us understand the sky as it appeared at the time of writing.

Was Shakespeare interested in the night sky? The frequency and depth of his references make it virtually impossible to argue that he was not. In attempting to interpret these references, I suggest contexts that complement more traditional approaches; in some cases it is even possible to point to a specific sky event, like an appearance of Venus in the evening sky that follows or precedes the accepted date of composition. Besides helping to interpret the works themselves, these interpretations are designed to nurture comparisons between literary passages and night sky objects or events.

Compilatio can be seen as a creative act of interpretation. As annotative as it appears, it is more than a passive collection of historical and literary facts, but is instead a recompilation of a series of experiences that the writers of that distant time shared with their readers.

Spiritus Mundi

> From four centuries in the future comes a hazy image of
> A shape with lion body and the head of a man,
> A gaze blank and pitiless as the sun …
> (Yeats, "The Second Coming", ll. 14–16).

The idea of *Spiritus Mundi* emerging from Yeats's "The Second Coming" is intended as a "universal subconscious" or source of meaningful images or poetry. Although it is a twentieth century term, the idea is forged from the past, where common images used in plays and poetry have a shared provenance. The eclipses and storm in *King Lear* (1.2.101), the star "westward from the pole" at the opening of *Hamlet* (1.1.35–38), the image of Romeo cut into little stars in *Romeo and Juliet*

(3.3.21–25), and even the "rotten humidity" of *Timon of Athens* (4.2.1–2) are all natural images apparently emerging from nature's grand repository. They are not personal, nor are they specific to a single writer; they appear to be available to anyone with the temerity to explore the natural world. Shakespeare, Spenser, Sidney, Jonson, and Fletcher all possessed this skill, as did writers in later eras, like Wordsworth, Byron, Keats, Tennyson, Hopkins, and Yeats.

Sky-Related Writings at This Time

It seems likely that specific major events in the sky, like blazing stars and eclipses, generated an increased interest in other, less important events in the sky. Sidney's *Countess of Pembroke's Arcadia*, completed around 1580, contains Klaius's hymn to the planet Mercury as seen at different times in the evening and in the morning (Sidney, *OA*.4.7–10). The celestial clock is used in the literature of the time, and examples of it, as well as their significance, are part of this dissertation. One particular example is Spenser's use of lunar phases as a clock in *The Faerie Queene*:

> Now haue three Moones with borrow'd brothers light,
> Thrice shined faire, and thrice seem'd dim and wan (*FQ* 3.3.16); again in Book 4: "But till the horned moone three courses did expire" (4.6.43); and once more in Book 5: "As the faire Moone in her most full aspect" (5.5.3). Since Virgo rises at dawn only at one particular time in the year, it is useful as an indicator of the time of year of a particular incident. The Moon's orbit around the Earth, on the other hand, is a timepiece that measures either the passage of months (three courses) or the particular time of a month (most full aspect).

(C) Discussion.

Although general discussions of the use of astronomical references in Shakespeare and his contemporaries have appeared from time to time, it is far less common to find investigations that emphasize the literary reaction to actual events in the sky as opposed to theories about it. Discussion about astronomy in Shakespeare seems to occur in cycles; in the 1930s a series of letters to the *Times Literary Supplement* considered the eclipses in King Lear, but the subject has not been considered actively since then despite the recent advent of far more accurate maps showing precisely how the eclipses were viewed from London. One of the *TLS* letters, for example, diminishes the importance of the October 1605 solar eclipse as a small partial eclipse (Harrison et al. 836, 78, 96). We now know (Espenak) that the event was a major eclipse in which almost the entire Sun was covered by the Moon, and that the lunar eclipse that preceded it was a deep partial eclipse which, at maximum, covered the entire Moon in either the Earth's umbral or penumbral shadow. A new look at these early modern celestial events would benefit from the more precise models that are available today.

In his 1922 study on astronomy and poetry, Cumberland Clark's *Astronomy in the Poets* does go beyond traditional references to stars and astral influence, pointing out less obvious but important references to the night sky. "Galileo read the open volume of the sky," wrote Clark, "while Shakespeare described its beauties to enrich his verse." These references include some of the celestial moods that are so evocative, including the many descriptions of dawn and sunrise that appear throughout Shakespeare (*Hamlet* 1.1.166–167, *Romeo and Juliet* 1.1.116, 130–134, 2.1.1–6, *Venus and Adonis* 1–2, 856). These poetic allusions cover a broad range of seasons and weather patterns, and offer the casual reader a way to relate to the sky of the time.

If Bacon's 1605 treatise *The Advancement of Learning* was a call to arms to learn about nature through direct observation, the invention of the telescope, and its use as a tool to increase humanity's appreciation of nature, fortified that call. The use of the telescope to study the sky, and its immediate consequences during the following decade, are events that delimit the terminal point of this study. This book expands on the general references provided by reviewers like Albanese, who painted a picture of humanity's appreciation of the sky from that time and how it led to the creation of new interests in extant forms of literature.

Studies about conditions, appearances, or events in the sky as described in the literature of the time seem relatively sparse in comparison to research about the astrology of the time. Garin's *Astrology in the Renaissance*, Grant's *Planets, Stars, and Orbs,* Meadows' *The High Firmament*, Russell's *The Copernican System in Great Britain*, and Alan Weber's dissertation *Shakespeare's Cosmology* are among the sources available. "Shakespeare—for or against astrology" (Whitfield 178–79) notes that Shakespeare, though not a fatalist, suggested that "the rise and fall of great men was subject to some external power." Whether he actually believed that is unknown, but what is important is that he respected the fact that his audiences did. The result is a richer collection of allusions to the night sky.

In relating the astronomical events to the literary works of this period, I intend to use both astronomical and literary sources. Astronomical sources will include works dating from both recent and Shakespearean times, plus letters, diary entries, and other evidence that astronomical events taking place in the early years of the seventeenth century were noticed and commented upon by the population, from James I down to his lowliest subjects.

(D) Basic plan. This section outlines the six chapters of this dissertation:

Preface
General Introduction
(1) The *stella novae* of 1572 and 1604
(2) A rich harvest of comets from 1576 to 1607, as well as possible meteor showers and storms around the turn of the century
(3) The eclipses
(4) The "great conjunctions" of 1583 and 1603
(5) The telescope
Appendix: A selection of references to the sky in writings from 1572 to 1620
Conclusion
Works Cited

(E) *Criteria for identification of astronomical references or allusions*. My method for identifying astronomical references or allusions will subject them to the following criteria:

1. Can the reference be connected to:
   (a) a specific event in the sky? or
   (b) a kind of event in the sky (like an eclipse) with which readers would be familiar? or
   (c) a more general cultural belief about the sky?
2. How can the reference help us to understand an aspect of the contemporary night sky?
3. Can the reference help us to appreciate the author's intent within a particular passage?
4. How does the reference function within the context of the writing?

A Personal Note

If one must choose a different sky from that of our era to pursue, why this particular one? It is the height of good fortune that two supernova events—those of 1572 and 1604—and a string of comets dotted the sky during the same period that produced the works of Shakespeare and Marlowe, even more so considering there had not been a similar nova since 1054, nor has one been so brightly visible in our own Milky Way galaxy since then; that of 1181 was less widely observed, with only Asian mentions in the historical record. What was the effect that these stars had on contemporary creative literature? In 1605, the year of the three eclipses, the 1604 supernova in Ophiuchus was still bright in the night sky, and Galileo's first use of the telescope was but 5 years away. Who would the observing partners of the time have been? I had always thought that Francis Bacon would have been one of them, but he spent almost of all the leisurely months before Parliament opened on 5 November 1605 (in the shadow of Guy Fawkes) writing *The Advancement of Learning*. Considering that Bacon is almost universally credited with developing the scientific method with its emphasis on rigorous observation and experimentation, I was surprised to find no reference to the rare series of eclipses during the year of its writing, or to the 1604 supernova, which both offered evidence of knowledge that could be gleaned from direct observation; instead I found only Bacon's general statement that "The astronomer hath his predictions, as of conjunctions, aspects, eclipses, and the like" (Bacon 112).

Other authors in 1605 did, however, refer to these events. While Samuel Rowlands joked over how seriously some people took the unusual astronomical events, I suspect he would have made an excellent observing partner in those last exciting years between the supernova of 1572 and the first turning of a telescope to the sky in 1610:

> His dinner he will not presume to take
> Ere he aske counsell of Almanacke
> Perhaps he spake it when the Moone did change
> And thereupon no doubt th'occasion sprung
> Unconstant *Luna* over-rul'd his tongue.
> Astronomers that traffique with the skie
> By common censure sometime meet the lie;
> Although, indeed, the blame is not so much
> Where Stars and Planets fail, and keep not tutch. (Rowlands, n.p.)

By taking this relatively new interdisciplinary field as its topic, this dissertation departs from deep interpretation in a narrow branch of English literature to a more overarching set of insights spread over a broad segment. It is my hope that these pages will help inspire students to explore more fully these ideas.

Today we are still debating the relationship that humanity has with the cosmos, though at a different level. Eclipses do not affect our futures, to be sure, but comets colliding with the Earth might well have brought the building blocks of life—the carbon, hydrogen, oxygen, and nitrogen that have been called the "simple alphabet of life." If that is true, then humanity indeed has extraterrestrial origins. As a practitioner of science in a time of deeply rooted beliefs in the power of cosmic forces, Kepler was a pivotal figure in defining the role of astrology at that time; he prepared horoscopes at the same time that his observations of the supernova of 1604 and his development of the laws of planetary motion were being taken seriously throughout England and Europe. Kepler accepted the prevailing ideas of judicial astrology and the idea that some events had metaphysical causes. His vision of astrology, however, had less room for the idea of signs than for the simplicity and elegance of the solar system (North 313, 318). He also objected strongly to the spiritualism and demonic magic that was practiced at the time by John Dee (Casaubon 22, Woolley 1) and by Lewes Lauaterus (83) in his 1572 book *Of Ghosts and Spirits Walking by Night.*

This is the study of a time, not of a specific author; it is an investigation of the relationship of two fields of endeavor, and not of a single aspect of literature. It focuses on how a series of actual events in the sky, rather than theories about the Universe, was an important motivating factor in the creation of new literature, and explores these celestial events as they are described, alluded to, or imagined, in different texts by different authors. The extent to which there are accurate representations of celestial phenomena in early modern English Literature has hitherto been underestimated (Figs. 1 and 2).

Fig. 2 The Bloomberg Library of the Hebrew University of Jerusalem, where the author received his Ph.D. in 2010, is one of the best research libraries in the world. Photograph by the author

Part I
The Sky in Early Modern English Literature

Chapter 1
The Stella Novae of 1572 and 1604

A star, a daystar, a firedrake rose at his birth. It shone by day in the heavens alone, brighter than Venus in the night, and by night it shone over delta in Cassiopeia, the recumbent constellation which is the signature of (Shakespeare's) initial among the stars. His eyes watched it, lowlying on the horizon, eastward of the bear, as he walked by the slumberous summer fields at midnight … (Joyce 210).

On one November evening in 1572, could a Stratford father and his impressionable 8-year-old son stood outside their home looking incredulously at a brilliant star just atop the "W" shaped figure of Cassiopeia? Far brighter than any other, this "blazing starre" shone even in daylight. While he left no documentation of his observation, it would have been hard for Shakespeare to miss it; even the following summer the star would have shone brightly low in the northeastern sky, "eastward of the bear" and low in the sky. The idea that the new star might have left an impression on him may be traced to a suggestion by Donald Olsen that Shakespeare had the nova in mind when he wrote these lines at the opening of Hamlet:

> Last night of all, When yond same star that's westward from the pole
> Had made his course that part of heaven Where it now burns, Marcellus
> and myself, The bell then beating one … (*Ham*.1.1.37–39).

There is no bright "Yond same star" (1.1.36) between the pole and the western horizon on late November evenings. Joyce expressed a similar idea in *Ulysses* though he did not relate it directly to the lines in *Hamlet*, suggesting that a youthful Shakespeare might have seen the supernova of 1572 "as he walked in the slumberous summer fields at midnight" (Joyce 210). Although further evidence that Shakespeare had this star in mind is probably not forthcoming, it seems plausible.

If we take a literal interpretation of the weather descriptions in the opening of *Hamlet*, we can infer that this scene is set on a frigid winter night. Hamlet's father had been dead for at least 2 months, killed while napping outdoors presumably on a

D.H. Levy, *The Starlight Night*, Astrophysics and Space Science Library 419,
DOI 10.1007/978-3-319-19878-1_1

warm afternoon, but this night is "bitter cold" (1.1.8). If Hamlet's father lost his life in late summer, and if the play opens at 1 am on a cold night, that night could well have fallen in winter, perhaps in January when the constellation of Cassiopeia would have appeared between Polaris, the pole star, and the western horizon. As noted by Olson *et al. (68–73)*, no star lies westward from the pole; all stars are south of it; however Shakespeare was probably thinking of a star between the pole and the western horizon (*Sky & Telescope*, November 1998, 68). Although it can be argued that Shakespeare had no particular star in mind, the supernova of 1572 happens to fit remarkably well within the parameters of Bernardo's speech if the event took place in midwinter. There is certainly much evidence throughout the canon that Shakespeare's great curiosity about the natural world extended to the sky, and it seems likely that he would have followed the star as it slowly faded during the 2 years it was visible.

Literature Preceding and After the Supernova's Appearance

In February 1570, astronomer and advisor to Queen Elizabeth I John Dee published his Preface to The *Elements of Geometrie of the most ancient Philosopher Euclide of Megara*. It outlines the shape of the sky as it was understood at the time:

The *Sonne*, when he is fardest from the earth (which now, in our age, is when he is in the 8.degree, of Cancer) is, 1179 semidiameters of the Earth, distante. And the *Mone* when she is fardest from the earth, is 68 Semidiameters of the earth and 1.3. The nearest, that the *Mone* cometh to the earth, is Semidiameters 52 ¼. The distance of the Starry Skie is, fró vs, in Semidiameters of the earth 20081 1/8(illegible). Twenty thousand fourscore, one, and almost a halfe. So thicke is the heauenly Palace, that the *Planetes* haue all their exerciuse in, and most meruailoufly perfourme the Commandement and charge to them giuen by the omnipotent Maiestie of the king of kings. This is what, which in Genesis is called *Ha Rakia*. Consider it well. (Dee, b2.)

The "blazing starre" was bright enough to be seen in daylight. On its opening night, 6 November 1572, it was first recorded at Horatio's Wittenberg (Olsen 68). Tycho Brahe used the sextant he had just built to learn how to measure star positions, including that of the new star, whose unchanging position proved to him that it was part of the "eighth sphere" of fixed stars (Brahe 3–4, Dreyer 38–41, Thoren 58). His measurements, showing that the star, unlike a comet, never moved among the stars, demonstrated that change could occur in that sphere (Christianson 17). The appearance of this new star was especially important because it called attention to the idea that the eighth sphere might not be so fixed as had been thought. This was likely of great interest to Queen Elizabeth I herself, since her own close friend and science advisor John Dee wrote a pamphlet about it (Dee, 1572).

When Tycho, along with anyone else who looked up at the sky, first saw the supernova, it was at least as bright as Venus, bright enough that it could be detected in daylight if one knew where to look. It remained visible throughout 1573 until

February 1574. But in the literary world it remained visible for years as writers tried to connect its significance to their lives and those of their readers. This literature ranged from direct attempts, both scientific and theological, to understand the star's significance to references in poetry and different forms of prose, including satire.

Even though some literary critics define a "blazing star" as being either a comet or a nova, it seems that the difference in appearance was well known in this time. Hamilton (308) defines blazing starre as a comet in *Faerie Queene* (3.i.16) because of the context of the object's "hearie beams and flaming lockes"; however both Shakespeare and Jonson refer separately to comets (*JC*. 2.2.30) and stars (*Volpone*. 2.1.50) as separate objects. The best known tract specifically about the new star was written by Tycho himself. *Learned Tico Brahe his Astronomicall Cometur of the new and much admired * Which appeared in the Year 1572* did not appear in English until 1632, but it was widely available in Latin at its publication in 1602 (Brahe). Its effect on English astronomical literature is offered by its translator "V.V.S.", who considered his achievement of bringing Tycho's new star to the wide English readership to be as significant as the star itself: Yet here the Reader shall most amply find

> Renowned Tycho's owne *Prognostication*
> Of the new *Starre* in this same new Translation …
> But wee make Tycho speake even word for word.
> Yet with that leave which * Horace doth afford,
> Who thinks it merits a Translators name
> To change the words, and yet the sense retaine,
> For this same Starre it selfe before did shroud
> Within the Latine, hid as in a Cloud,
> But now it is unvayl'd, and here in sight
> It shineth forth againe, as cleere and bright
> As when it first appeared in the *Skie*,
> And was the object of each wandering *Eye*. (Brahe, n.p.)

The translator's offering may not have been the best representation of contemporary English verse, but it correctly identifies the idea that Tycho's insights on the star would be available to a great variety of readers, including this brief "elegy" by Scotland's James VI the year before he ascended the throne of England as James I.

> What Phaeton dar'd, was by Apollo done
> Who rul'd the fiery horses of the Sunne.
> More Tycho doth; hee rules the Starres above
> And is Urania's favorite, and Love. (Brahe, n.p.)

Tycho wrote that the star of 1572 "shined without a taile or any scattered beams (for then it had beene a Comet) yet neverthelesse it might be likened to some of those appearances …" (Brahe, 4). Brahe correctly discerns that the star is "coelestiall, not differing from the matter of the other Starres, but yet in this it did admit of some diversitie, that it was not exalted to such a perfection … as appeareth in the everlasting and continuing Starres" (Brahe 9). Essentially, Tycho was saying that the star was the same, yet not the same, as the other stars; this was a good way towards our present day understanding that the object was indeed a massive star at the very end of its life. Throughout his work he was also using the "numbers" of astronomical and astrological calculation to depict the star's position in the sky and the possible influence of that star on the Earth's northern hemisphere. Tycho's discussion of the possible astrological effects of the star was somewhat reticent: "I think it not fit," he concluded, "to proceed any further in unfolding these mysteries, having promised to handle them sparingly; also in regard, that these Prophesies, are not to be declared by humaine conjecture, neither can be Geomertically demonstrated, as those matters which belong to the knowledge of Astronomie." Although Tycho's words were not published in English until 1632, they were well known in their original Latin in time for Shakespeare to consider them in Hamlet's letter to Ophelia:

> Doubt that the stars are fire;
> Doubt that the sun doth move;
> Doubt truth to be a liar;
> But never doubt I love
> O dear Ophelia, I am ill at these numbers; I have not art to reckon my groans: but that I love thee best, O most best, believe it. Adieu (*Ham*. 2.2.115–121).

Shakespeare was possibly familiar with the words of the scientist whose name was tied to the great star of 1572, and whose work inspired the elegy by King James. Thus the words in *Hamlet* may well have been inspired from Tycho. There is some evidence that Shakespeare was aware of the details of Tycho Brahe's life and family. Olson (68) points out that Tycho's observatory at Uraniborg was not far from Hamlet's castle in Elsinore; on a clear day Tycho could look out across the water and see it. One of Tycho's books, printed at his observatory in 1596 and titled *Epistolarum astronomicarum*, boasted a portrait of Tycho surrounded by the coats-of-arms of his ancestors, which included Rosenkrans and Guldensteren, names nearly identical with those of Hamlet's two fellow-students hired by Claudius to snoop on Hamlet. Shakespeare might have seen the book or borrowed it from Digges, and possibly he enjoyed the unusual names enough to make use of them (Gingerich 394–95).

The authors of one of Shakespeare's most prominent sources, *Holinshed's Chronicles*, wrote of the star in his narrative of the events of 1572 The eighteenth of November in the morning was seene a star northward verie bright and cleere, in the constellation of Cassiopeia … appearing bigger than Jupiter, and not much lesse than Venus when she seemeth greatest. Also the said starre never changing his place, was carried about with the daily motion of heaven, as all fixed stares commonlie are…. (Olson 70, Holinshed 1257)

The *Chronicles* properly note that the star was brighter than Jupiter and almost as bright as Venus at greatest brilliancy. They also record correctly that Venus varies in magnitude depending upon its distance from Earth and its phase.

Other writers noted the great star. In 1580, Francis Shakelton's pamphlet *A blazyng Starre or Burning Beacon* recorded that "On the 18th day of November in the morning, a Starre was sene Northward, verie bright and cleare in the constellation of Cassiopeia, and it seemed so strange" (Shakelton D3).

Other writers described the wonderful star in Cassiopeia; In his historical log of English events to 1604, Harison describes:

> The 18. of November was seene at that Northward very bright and cleare in the constellation Cassiopeia, which with three chiefe fixed stars of the said constellation, made a geometricall figure … This starre in bigness at the first appearing seemed bigger than Jupiter, and much less than Venus when she seemed greatest, also the said starre never changing his place, was carried about with the daily motion of heaven as all fixed starres commonly are, and so, continued almost five months. (Harison, 302)

The supernova of 1572 which Harison clearly describes as being about magnitude −3, and the following one in 1604, only 32 years later, were remembered by writers who were not yet alive at the time of its appearance. (There have been no supernovae observed in our Milky Way galaxy since then.) In 1647 Henry More posed questions that reflected popular concerns of the time. How did a new star manage to appear in a sky full of ordinary, permanently placed stars, and then how did it disappear; where did it go?

> And new fixt stares found in that Circle blue,
> The one espide in glittering Cassiopie,
> The other near to Ophiuchus' thigh,
> Both bigger than the biggest stares that are,
> And yet as farre remov'd from mortall eye
> As are the farthest, so those Arts declare
> Unto whose reaching sight Heavens mysteries lie bare. (More 210)

Indirect References to the 1572 Supernova

A search for possible indirect references to the star of 1572 can provide some insight to its connection to human thought and perception, particularly concerning the journey of the soul after death. In Spenser's account of the battle between Cambell and the three brothers in *Faerie Queene*, the soul after death is given three normal choices (*FQ* 4.3.13, Hamilton 445), one of which is to change "into a starre in sky." But no nova brightens with this death, as the soul moves into a brother to continue the battle. In *Nosce Teipsum*, Davies echoes the role of the sky in the opening explanation to the soul's progress after death: "The lights of heav'n (which are the World's faireies) Looke down into the World," and in 'An Acclamation' the soul appears with starlike beams. Donne's *First Anniversary* describes how "in these constellations then arise: New starres, and old do vanish from our eyes." (*First Ann.* 260)

New stars appear in Christopher Marlowe's works as well. "Why addest thou stars to heaven, leaves to green woods? And top the vast deep sea fresh water floods?" Marlowe might have recalled the bright star in Cassiopeia as he was translating Ovid's elegies (2.10.13). In this elegy to Graecinus, Ovid (as translated by Marlowe) treats temporary stars in the sky as casually as leaves growing on trees; the transience of the great new star of 1572 was obvious by the time of the appearance of Marlowe's translation in 1594, for although it never changed position, by then it had completely vanished from view.

Marlowe wrote *Tamburlaine* around 1587 about the glories in battle and death of a fourteenth century Scythian shepherd who became a great conqueror. Marlowe's Tamburlaine wishes that

"Over my zenith hang a blazing star,
That may endure till heaven be dissolv'd
Fed with the fresh supply of earthly dregs,
Threat'ning a death and famine to this land!" (*2 Tamburlaine* 3.2.6–9)

The idea that this particular blazing star endure forever indicates that Marlowe is thinking of a star apparently fixed to a spot in the heavens relative to other stars, rather than a comet. That he probably was thinking of the Supernova of 1572, rather than a comet as some writers take blazing stars to mean, is evidenced by his specific invocation of different types of objects and phenomena: He considers "fiery meteors" (3.2.4) and two acts later he writes of "comets and blazing stars" as different types of heavenly objects. In the play's final act Marlowe describes the "sturdy Governor of Babylon" about to go to his death by hanging upon the ruins of the town he rules. "Go, bind the villain!" Tamburlaine orders amidst a view of a battlefield more terrifying than if the skies

Were full of comets and of blazing stars
Whose flaming trains should reach down to Earth
Could not affright you. No, nor I myself,
The wrathful messenger of mighty Jove,
That with his sword hath quail'd all earthly kings …
(*2 Tamburlaine* 5.1.89)

Tamburlaine can be considered an early example of science fiction, a play about a foreign monster growing up peaceably in the towns of Renaissance England. It is science fiction in the sense that the lifestyles, morals and even the spirit of the plays are set in a place thousands of miles away from the play's audience, an audience that was facing a new world whose creatures were, as Kirschbaum notes, "utterly abnormal emotionally, ethically, and religiously to the citizens of Shakespeare's London" (Kirschbaum 28–29).

Just as in our time, the appearance of an unusual star or comet triggered an increased interest in other, less important events in the sky. Sidney's *Countess of Pembroke's Arcadia*, completed around 1580, contains Klaius's hymn to the planet Mercury as seen at different times in the evening and in the morning sky:

O Mercury, foregoer to the evening;
O heavenly huntress of the savage mountains;
O lovely star, entitled of the morning;
Vouchsafe your silent ears to plaining music,
Which oft hath echo tired in secret forests.
(*Old Arcadia Fourth Eclogues*)

The hymn has increased meaning to readers who are aware of the changing positions of the planet Mercury as a result of its orbit around the Sun. Since Mercury is closer to the Sun than Earth, whenever it appears in the sky it is near to the horizon in either evening or morning twilight. Occasionally, Mercury climbs higher in the sky and becomes more easily visible, as it did on 3 November 1579, in the morning sky and again in the evening sky 21 June 1580 (Mercury Chaser's Calculator). Both Mercury and Venus appear alternately in the evening sky for several weeks, then not at all as they get too close to the Sun to be seen; then each appears for a new period in the morning sky. Rarely during that middle period of invisibility, Mercury and Venus can actually pass in front of the Sun in an event called a transit; the most recent transit of Mercury occurred on 8 November 2006.

The Star and the Harvey-Nashe Controversy

Buried in the rich body of early modern literature of the late Elizabethan period is a collection of satirical pamphlets by Thomas Nashe and other writers. Not commonly read today, their lively styles and insulting tones were popular in contemporary England, reaching an ever-widening readership with each successively nasty creation. Originally written by an author using the pseudonym Martin Marprelate, the pamphlets began as a criticism of Archbishop John Whitgift's 1586 decree banning the publication of any book or pamphlet not authorized by him. In 1589, Nashe was hired by the government as a propagandist to defend the ban (Nicholl, 62). In probably the only paid writing job Nashe ever had, he had his first opportunity to earn his keep the following year, when Richard Harvey attacked the anti-Marprelate faction. A brilliant satirist, Nashe often used the night sky as a means of accenting the barbs he wished to throw, especially against those who are "so privy to the secrets of the Almighty that they should foretell the tokens of his wrath … no star he seeth in the night but seemeth a comet." (*Anatomy of Absurdity* 1589)

In 1592, Robert Greene's *A Quip for an Upstart Courtier* launched a subset of the Marprelate hullabaloo, a peculiar collection of nasty pamphlets tossed between Gabriel Harvey and Thomas Nashe. (Greene apparently liked using the word "upstart"; history remembers his calling Shakespeare an "upstart crow" in his *Greene's Groatsworth of Wit*). *Upstart Courtier* talks of those arrogant courtiers—possibly including Gabriel Harvey—who "lifted their heads so high as if they had been bred to look no lower than stars". After Greene's death that year, Gabriel Harvey and Thomas Nashe launched their own set of pamphlets directed against each other, a conflict that became personal and bitter, and took up the rest of their lives. Harvey's last letter of *Foure Letters, and certaine sonnets* offers friendship to Nashe in exchange for ending the feud and ceasing his hotheaded style. Harvey alludes generally to the sky, first to the *ignis fatuus*, or mirage, of a "sky never so spritishly busy; never so many threatening comets, never such a terrible sky of blazing and falling stars" and too, in sonnet 11, the "resplendent lights of Milky Way to sing." Nashe's response is the pamphlet *Strange Newes* (1593) to which Gabriel Harvey replied with *Pierce's Supererogation, or A New Praise of the Old Ass* (1593) the attack on his nemesis gets more personal. His fame, which can be inferred not in a positive light to mean evil reputation (OED), is given a persona: "When his necessary defence hath sufficiently accleared him whom it principally concerneth to acquit himself, she (Saint Fame) shall no sooner appear in person, like a new star in Cassiopeia, but every eye of capacity will see a conspicuous difference between her, and other, matters of eloquence, and the woeful slave of St. Fame must either blindfold himself with insensible perversity, or behold his own notorious folly with most shameful shame." Harvey is obviously accusing Nashe of having an ego as big as a supernova.

Nashe gets his comeuppance in his 1596 pamphlet, *Have with you to Saffron-Walden* (4535), a satire about a fictional imbecile who substitutes for the real Harvey, and in which Nashe accuses Harvey of behaving almost maniacally. *Saffron-Walden* marks the pinnacle of Nashe's battle with Harvey, who lived some 15 miles from Cambridge in a town called Saffron Walden. In it Nashe notes the character's "hateful scribbling" about a number of natural events, including "the fearful blazing star." Nashe even revisits the supernova issue as part of a list of how he referred to Saint Fame: "In one place he (Gabriel Harvey) calls her (the "sweet gentlewoman") Saint Fame *the one she*, in another *the credible gentlewoman*, in a third *the heavenly plant*, and the fourth *a new star in Cassiopeia*, in the fifth *the heavenly creature*, in the sixth *a lion in the field of Minerva* …" and later compares Harvey and his gentlewoman to Jupiter and the Moon.

Saffron-Walden was one of the books ordered destroyed, and never reprinted, by the archbishop of Canterbury in 1599 in an attempt to put the literary skirmish between the two writers to rest. Even though the star had appeared 24 years earlier and had long since faded away in the sky, it lived on as a starry culmination to this footnote of early modern literary history. However, the supernova allusions appear as a doorway to a rich body of celestial allusion. The night sky, and particularly the attempts of judicial astrologers to interpret it in artificial ways, offered a primary source of fodder for Nashe's satirical pen. The more one delves into this material, the better one appreciates the depth to which Nashe has done his homework.

In *Summer's Last Will and Testament*, Nashe refers obliquely to Tycho Brahe, whose precise mathematical measurements of the sky led to his determination that the supernova of 1572 belonged to the most distant celestial sphere.

> Sky-measuring mathematicians,
> Gold-breathing alchemists we also have,
> Both which are subtle witty humourists,
> That get their meals by telling miracles.
> Which they have seen in travelling the skies;
> Yalm boasters, liars, make-shifts, they are all,
> Men that, removed from their inkhorn terms,
> Bring forth no action worthy of their bread. (Nashe, 189)

In his blanket condemnation of all judicial astrologers and alchemists, Nashe includes Tycho, whose *Astronomicall Cometur* on the supernova paraded some ideas about the supernova's sudden appearance foretelling the end of the Earth. To use the material of astrology as a basis for satire, Nashe needed to know it thoroughly, from true soothsaying to the borderline prognostications of Tycho, as part of his otherwise scientific treatise.

Tycho did indeed consider the star's judicial astrological implications at the end of his epistle: "As this Starre appeared in the highest heavens, to the view of the whole world, so it is credible, that there shall happen a great Catastrophe and universall change throughout all the chief Nations of the Earth, especially those which are situated Northward from the Equinoctiall" (Brahe, 16). His discussion then turns to the possibility that the star is somehow related to the Great Conjunction coming in 1583 between Jupiter and Saturn, an apparent meeting of planets in the sky that occurs once every 20 years. It is a celestial event of great interest and which Shakespeare wrote about in *All's Well that Ends Well*, as detailed in Chap. 4.

The new star of 1572 was not just an unusual object in the sky. As historian William Camden expressed in his *Annales*, "in the month of November a new Starre, or if you will, a *Phaenomenon*, was seene in the Constallation of *Cassiopeia*, which (as I my selfe observed) in brightness excelled Jupiter …" (Olson, 70). A phenomenon indeed, this star's appearance and ramifications affected, as I have tried to show by these examples, the writing of a generation. Although we will never know how the star might have inspired the 8-year-old Shakespeare, it is surely possible that a youthful view of a strange star in the night sky could have inspired him, and other writers, to look beyond daily life to develop a cosmic vision for their time (Figs. 1.1, 1.2, and 1.3).

Fig. 1.1 This image shows two towers standing prominently over the Edgbaston waterworks Towers in Birmingham, England. Although these genuine towers might have inspired J.R.R. Tolkien's *The Two Towers* (the second installment of his famous trilogy *The Lord of The Rings*), here I imagine the scene several centuries earlier, during the great solar eclipse of October 1605 during which more than 85 % of the Sun was covered by the Moon from Birmingham; about 90 % from London

Fig. 1.2 NGC6946: NGC 6946, a galaxy 23 million light years away. It is best observed in summer and fall in the northern hemisphere. There are no supernova in this galaxy at present, but exploding stars frequently appear here, as in 1917, 1939, 1948, 1968, 1969, 1980, 2002, 2004, and 2008. Photo by Wendee Wallach-Levy. This picture served as the back cover photograph for the Royal Astronomical Society of Canada's Observer's Handbook for 2010

Fig. 1.3 SN 1054: The bright "star" in the center of this photograph is in almost the precise position of the great supernova of 1054 in our own Milky Way galaxy, but it is not the supernova of 1054. Instead it is Saturn that just happens to be very close to where the real supernova was all those centuries ago. This photograph, taken by the author, offers an impression of what a dark sky might have looked like over a thousand years ago

Chapter 2
Comets and Meteors: A Rich Harvest from 1573 to 1607

When I discovered a comet on the morning of 14 June 1991, I had no idea that this "brave new world" (*Tempest*.5.1.183–184) would actually take me on a cometary journey into the sky of a distant time and place. As the comet's motion was documented over the next weeks, it was possible to determine its orbit over many years. In the past, it turned out that this comet might have visited us before, and it could be the same body that appeared in September 1499 (IAUC 5306, 1991 13 July). Near that ancient time the Italian poet Jovianus Pontanus was writing *Urania*, a Latin work about the night sky written in hexameters, that he would read at an Academy meeting in 1501. Pontanus's words appeared again in Latin around the middle of the sixteenth century, as part of Friedrich Nausea's *A Treatise of Blazing Starres in Generall*. The small portion of Pontanus's work that relates to comets was translated into English by Abraham Fleming at the end of the fifteenth century. Thus, the comet of 1499 might have encouraged Pontanus to write about comets. When his words appeared in English a century later, they might have been inspired by a spectacular parade of 15 comets (Yeomans, 1991) that appeared between 1573 and 1607. This increased frequency of comet apparitions is important for literary reasons, since references to them abounded in the writings of the time:

> Of whistling winds, with blustering blasts which blow,
>
> Of bloody broyles, by force in fatall sight:
>
> Of peoples pompe, the pitious overthrow.
>
> Of Potentates the death, in wofull plight:
>
> The Blazing Starres aloft like lamps of light
>
> In th' Est or West of azure coloured skies,
>
> Forewarnings and signes when they arise.
>
> If still they stand, and offer not to change
>
> The place, where first in sight they cast their beames,

D.H. Levy, *The Starlight Night*, Astrophysics and Space Science Library 419,
DOI 10.1007/978-3-319-19878-1_2

Fig. 2.1 For by these blessed candles of the night… (*Merchant of Venice* 5.1.220)
This recent comet, Comet Hyakutake (C/1996 B2), could represent a comet that might have been visible in Shakespeare's time

Then shall insue much mischiefe rare and strange:
As gaping wounds, and sluzing bloody streams,
In foughten field, twixt nigh adioyning Realmes,
Such civill stormes shall overrunne the land,
That some shall bath in kindreds blood their hand.
And if they take their course unto the East,

A signe it is that forreigne foes with force
Prepare apace, to spoyle both most and least:
With edge of sword sweet life still to divorce
From panting heart, devoyde of all remorse.
Has East or West in sight, they have their dome,
And signes they are of things in time to come. (Fleming 1.21)

Brought to life for the English speaking world by Fleming, these words depict the passage of a "Shakespearean" series of comets that began at the same time that Tycho's great star was still bright. The interpretation is astrological; "blazing stares" appear as "signes" alongside "stirring winds." A comet in 1573, and a second appearing in 1576, were followed by the great comet of 1577, another in 1580, yet another in 1581, and two more in 1582.

Other comets disturbed the order of the sky in 1585, 1587, 1590, 1591, 1592, 1593, 1596, 1600, 1601, 1602, and 1607—11 over 22 years (see Table 2.1). A more typical frequency for comets visible without a telescope is found during the same period of our own time, during which only seven bright comets appeared, in 1974, 1976, 1983, 1986, 1990, 1996, and 1997. Except for a remarkable series of three comets in 1618, comets appeared far less frequently after 1607 (Yeomans 414–18).

This proliferation before telescope observing was possible is extraordinary, and it is hardly surprising that Fleming chose to translate "A Prognostication of Blazing Starres, according to the opinion of the Poet Pontanus." Especially noteworthy is the poet's emphasis that "blazing starres" are either stationary or moving, an indication that Fleming wished his readers to understand that the new stars (i.e. supernova) and the comets had very different appearances and should be interpreted differently. The poem begins with the "stirring winds" and the blasts of storms that forecast "civill stormes" just as *King Lear's* storm scene is preceded by a series of eclipses that presage, as Gloucester suggests, "in palaces, treason … " (*KL* 1.2.103).

Stanza three considers the direction of cometary motion. The series of events that sparked Fleming's translation offered two stars "still they stand, and offer not to change," (possibly the new stars of 1572 and 1604) plus a selection of comets that "take their course unto the East" and others that could appear "East or West in sight, they have their dome, And signes they are of things in time to come." If they appear in the evening sky, then comets will be in the west and will "take their course unto the east" (move eastward) as they depart the vicinity of the Sun. As the sky began to darken after sunset on the night of 13 November 1577, Tycho Brahe, already famous for his observations of the new star of 1572, saw what appeared like a long, white, softly luminous cloud in the northwestern sky. Intrigued, he noticed that it brightened as it neared the western horizon. As the sky grew darker, it quickly became obvious that this was no cloud but the tail of the mightiest comet he, or anyone else alive at the time, had ever seen. According to J. Harison, "The 11. of November at night was seen a blazing starre with a long (train) which was nightly seene till eight of the clocke" (Harison, 324).

Table 2.1 Comets and Cometary references in Early Modern English Literature

Comet of	Literary Reference
1573	
1576	
1577	*A blazyng Starre or burnyng Beacon, Seene the 10. of October laste (and yet continewyng) set on fire by Gods providence, to call all sinners to earnest and speedie repentance*, [1578] compares the comet of 1577 to the view over Jerusalem 2500 years earlier: "No lesse than the Starre that stood over Jerusalem like unto a Sworde, and also the strange Comet which endured for the space of a whole yere: The Devill in the figure and shape of a man . . ." .
	Also in the yere of our Lorde 1577 the tenth daie of November at night," Shakelton added, "there appeared a blazyng Starre in the firmament with a long streame (or taile) proceadying from the Same, and stretching the beames thereof directly towards the Easte, verie wonderfulle to beholde." [Shakelton D4] Shakelton attaches these events to *Revelations* 6.13 and 14.7, where stars fall to Earth, and to the idea of darkening the Sun, Moon, and stars (8.12). (Shakelton [circa 1588], D4)
1579	Spenser's *Shepheardes Calender* (though possibly this harks further back to the comet of 1577, which was far brighter) stird vp that vnkindly heate, That reigned (as men sayd) in *Venus* seate." (Bainbridge, 1618)
1580	
1582	
1585	George Peele, *Araygnement of Paris* "The water-flowers and lilies on the banks . . . " Like blazing comets, burgeen all in ranks; Under the hawthorn and the poplar-tree Where sacred Phoebe may delight to be . . . " (Peele [1585])
1588	Nashe's *Anatomie of Absurditie* He falleth asleep; no star he seeth in the night but seemeth a Comet; he lighteth no sooner on a quagmire, but he thinketh this is the foretold earthquake, whereof his boy hath the ballad.
1590	First 3 books of *FQ*
1591	*1 Henry VI (although this reference may be generic to the several comets of the preceding years)* "Comets, importing change of times and states, Brandish your crystal tresses in the sky." (*I Henry VI*.1.1.2-3)
1592	frozen meteors *Summer's last will* [1592] I could bark the sun out of the sky, Turn moon and stars to frozen meteors And make the ocean a dry land of ice; . . . (*Summer's Last Will* 201) vapours before the Sun (*Pennilesse* [1592])
1593	
1594	Nashe *Terrors of the Night* brains are like meteors (*Terrors* 241[1594])

(continued)

Table 2.1 (continued)

Comet of	Literary Reference
1596	
1599	Shakespeare: *Julius Caesar* "When beggars die there are no comets seen, The heavens themselves blaze forth the death of princes." (2.2.29-30)
1600	circa 1600: Fleming
1601	
1602	

1. Halley's Comet.
0.My lord, that ne'er before invited eyes,
But have been gaz'd on like a comet."
(*Pericles* (5.1.83-85)

A cloud or a comet? Just as the fading light made the difference unclear at first, Tycho was aware that the prevailing theory of comets belonged to Aristotle, who opined that they were exhalations in the upper atmosphere not unlike ordinary clouds. Believing that this theory was wrong, Tycho made accurate observations of the comet's changing positions over time, and compared them with those made at the same time from other observers across Europe. When he discovered that positions made at the same time from different locations found the comet in the same place in the sky, he concluded that the comet had to be farther from Earth than the Moon is.

Although Tycho proved this point in 1577, John Bainbridge (in 1619, some months after the appearance of three comets in 1618) credited Seneca, in the *De Cometis* segment of his *Quaestiones Naturales*, with suspecting it fifteen hundred years earlier. "Let this Comets regular and ordinate motion (for a month together) never deviating from one right line, keepe you in the way of verity," he wrote "This very reason alone many ages agoe persuaded divine *Seneca* … to place all Comets above the elementary regions (meaning Earth's atmosphere), wherein only inconstant and momentany (sic) Meteors make their fickle vagaries." (Bainbridge 18) Bainbridge's science is correct, his attribution not entirely so. Seneca did not believe that comets were solar system objects beyond the moon; however he quoted Appollonius who did, and, thanks to the burning of the great library at Alexandria and the murder of Hypatia. *Qaestiones Naturales* is probably the only surviving source of Appollonius's view. (Seneca, *De Cometis*. XV11.1)

Look how a comet at the first appearing Draws all men's eyes with wonder to behold it …

Comets and Meteors in Shakespeare

Around 1590, Shakespeare opened 1 Henry VI with these lines from Bedford:

Comets, importing change of times and states,

Brandish your crystal tresses in the sky. (1.1.2–3)

In a single three-line passage Shakespeare invokes much that changes in the sky; the opening line implies a solar eclipse (see Chap. 3), where day involuntarily and suddenly yields to night, followed by a precession of comets. Note again the plural form; by 1590 Shakespeare might have seen as many as eight comets, using the word that derives from the Greek "long-haired star", and he decorates his comets with tresses like crystal. I doubt that the crystalline image could have entered Shakespeare's mind had he seen a comet only through the description of another, whether verbal or artistic. The words suggest that he saw more than one of the comets, perhaps including the mighty Tychonian comet of 1577 whose tail stretched a quarter of the way across the sky.

The bright comets continued through these years, especially the great summer comet of 1596 which might have been in Shakespeare's memory as he composed *Julius* Caesar around 1599. Calpurnia warns her husband not to venture outdoors after her dream of seeing comets in the night: "When beggars die there are no comets seen," Calpurnia warns her husband, "the heavens themselves blaze forth the death of princes" (2.2.29–30). Shakespeare's invocation of a comet in *Julius Caesar* is historically significant, for during the Octavian games in the summer of 44 BC, held in memory of Caesar, a bright comet tracked through the northern sky and was widely believed at the time to represent the Praetor's soul on its way to heaven.

The comets of Calpurnia's dream were premonitions that appeared before her husband's assassination, but the historical comet appeared 2 months afterwards, with a tail perhaps 12° long—half the length of the Big Dipper—moving through the northern sky, according to Plutarch; "among the divine portents there was also the great comet; it appeared very bright for seven nights after the murder of Caesar, then disappeared." Calpurnius Siculus went further, blaming the comet for the civil war that followed: "when, on the murder of Caesar, a comet pronounced fatal war for the wretched people."

> By being seldom seen, I could not stir,
>
> But like a comet, I was wond'red at;...
>
> . . . .
>
> That men would tell their children "This is he" … (*IHIV* 3.2.47–48)

If *I Henry IV* was completed by 1596 (Oxford 188), Shakespeare might have been thinking of the comet of that summer; the monarch and his subjects wonder at the majesty of a comet, but the allusion is inverted so that the King himself gets wondered at like a comet. The comet of 1596 was clearly visible in the northern sky throughout Europe for the first 2 weeks of August, seen first in Auriga, in the northeast, in mid-July and then moved eastward through Gemini and finally into northwestern Leo.

> Some airy devil hovers in the sky
>
> And pours down mischief. (*KJ*.2.2.2–3)

King John is particularly rich in cometary allusions. The *airy devil* implies a comet, since a quickly flashing meteor would not hover and airy summons the image of "a bearded star." If *King John* was completed in 1596 (Oxford, 240), then this line might have been inspired by the Great Comet of that summer.

Meteors

No natural exhalation in the sky,
No scope of nature, no distemper'd day,
No common wind, no customed event,
But they will pluck away his natural cause
And call them meteors, prodigies, and signs,
Abortives, presages, and tongues of heaven,
Plainly denouncing vengeance upon John. (*KJ*.3.4.153–159).

Meteors were thought of as "natural exhalations" and still are in a sense. They offer a rare example of a fundamental aspect of astronomy whose core has not changed much over four centuries; our current understanding of meteors is rooted in the idea that they are events within Earth's atmosphere. Each meteor is an event during which a speck of cometary dust heats to incandescence as it races through the upper atmosphere. We see only the interaction of the meteoroid, or dust speck, with Earth's atmosphere. To the early modern mind they could certainly have been seen as portents. But in *King John* these portents do not forecast vengeance. Meteors are here invoked by name and linked to the "prodigies and signs" that denounce, not pronounce, vengeance upon John, as if to suggest that John's actions, presumably done in his role as monarch and for the benefit of the people of England, will return to haunt his future:

This show'r, blown up by tempest of the soul,
Startles mine eyes and makes me more amaz'd
Than had I seen the vaulty top of heaven
Figur'd quite o'er with burning meteors,
Lift up thy brow, renowned Salisbury,
And with a great heart heave away this storm…. (*KJ*.5.2.50–55)

This allusion, in a speech by Louis the Dauphin of France, is not just to a single meteor but to a shower of meteors, and more likely a philosophical, *Revelations-type* shower where actual stars fall from heaven, than a physical one. However, it seems possible that Shakespeare witnessed, or more likely heard travelers' tales of a night during which several meteors per minute fell from the sky. The Leonid meteor shower of 1602, for instance, featured rates as high as several per second, but those were observed over China, not England. (ThienWen-Chih notes that On November 6, "hundreds of large and small stars flew, crossing each other"); the Korean text *unhon-Piko* notes that on November 11 of the same year, "many stars flew in all directions." We do not know if the shower lasted long enough to gain visibility worldwide, as did the Leonid meteors of 1998, which maintained high rates for a sufficient number of hours to be seen around the world. A similar idea is portrayed eloquently in the childrens' story *The Last Battle*, the final book of C. S. Lewis's *The Chronicles of Narnia*. (Lewis, 189)

Meteors fright the fixed stars of heaven;
The pale-faced moon looks bloody on the earth, (*RJ* 2.2.4.10–11)

Meteors may be portents here, but not as in *Revelations* since there is a differentiation between meteors and fixed stars; meteors are called upon to fright. Similarly,

Those oppos'd eyes,
Which, like meteors in a troubled heaven
All of one nature, of one substance bred,
Did lately meet in the intestine shock
And furious close of civil butchery (*1 HIV* 1.1.8–12)
Yond light is not daylight; I know it, I:
It is some meteor that the sun exhales
To be to thee this night a torch-bearer,
And light thee on thy way to Mantua; (*RJ* 3.5.12–15)

Here seen as an exhalation of the Sun, a meteor in Shakespeare is a portent, in this reference a natural "torch-bearer."

Now, see thy glory like a shooting star
Fall to the base earth from the firmament! (*R2* 2.4.19–20)

Meteor references in the early modern period allude to their astrological portents much more often than to celestial events; In *All is True* King Henry VIII predicts that

I shall fall
Like a bright exhalation in the evening,
And no man see me more. (*AIT* 3.2.225–227)

Seneca believed that comets are formed "by very dense air, and since the most sluggish air is in the north, they appear in greatest number" in that direction. Although comets are distributed across the sky almost at random, in Seneca's own experience the small-number statistics of comets happened to favor the sky in the north.

"Also in the yere of our Lorde 1577 the tenth daie of November at night," Shakelton added, "there appeared a blazyng Starre in the firmament with a long streame (or taile) proceadying from the Same, and stretching the beames thereof directly towards the Easte, verie wonderfulle to beholde" (Shakelton D4). Shakelton attaches these events to Revelation 6.13 and 14.7, where stars fall to Earth, and to the idea of darkening the Sun, Moon, and stars (8.12) (Shakelton D6).

Shakelton also references directly the Comet of 1580, as a further example of writing that recalls events actually seen in the sky: "And now hath the lord God kindled any other Starre, this late tenth of October 1580, the first appearing whereof (with his continuance) which hath been in Aquarius …" (Shakelton D5).

In 1579, the December portion of Spenser's *Shepheardes Calender* followed this initial string of five bright comets. As part of a reflection on life and death that defines a relationship between man and nature, Spenser invokes the appearance of a comet that, in the *Calendar's* argument for December, says "his manhoode to the sommer, which he sayth, was consumed with great heate and excessiue drouth caused through a comet or blasinge starre, by which he meaneth loue, whose passion is comenly compared to such flames and immoderate heate." The comet he meant was probably one that appeared in 1578, although he could have recalled also the comet of 1577:

> stird vp that vnkindly heate,
>
> That reigned (as men sayd) in *Venus* seate. (*Shep. Cal.* December 10.5–6)

Comets and Meteors of Our Time and Theirs

Comets can be the most stunning of celestial objects, as those of us who saw the spectacular sights of three bright comets in 1996, 1997, and 2007 can attest. In March, 1996, I observed Comet Hyakutake's ethereal, filmy tail stretching for 110°, almost across the entire visible sky. As the comet moved through space, gas and dust erupted from its nucleus. How would such cometary sights have been interpreted by the seventeenth-century English people in particular? Most important, the appearances of comets were so unusual that those who viewed them kept detailed records of their paths across the sky. Those records date back possibly to biblical times. The first book of *Chronicles*, for example, describes what could be a comet which might have been seen as a rebuke for an ill-advised census that King David had ordered. That biblical passage is read every year at the Passover Seder: "And David lifted up his eyes, and saw the angel of the Lord standing between the earth and the heaven, having a drawn sword in his hand stretched out over Jerusalem." (1 *Chr.* 21:16., cf. *And with an outstretched arm*: this refers to the sword, as it is said, "And a drawn sword in his hand outstretched over Jerusalem." (C. Roth, *Haggadah* 26))

If the end of David's reign occurred about 971 BCE, then two comets, recorded by Chinese sources, are candidates. During the war between two Chinese kings Wu-Wang and Chou, around 1059 B.C., a comet with an eastward-pointing tail dominated the evening sky. Another, appearing in the north polar region some time between 974 and 959 BCE, is within 5 years of an estimated date of the enumeration of 965 BCE. (Yeomans 362, Bible Timeline 1254 BC–1004 BC, Pinsky (2005, 2006), Ken Wade (2006)).

Shakelton defines a nova as being different from a comet although Spenser ignores that difference in *Faerie Queene* (3.1.16), where a blazing star is clearly a comet:

> All as a blazing starre doth farre outcast
>
> His hearie beams, and flaming locks dispred (*FQ* (1590) 3.1.16)
>
> And when faire *Cynthia*, in darksome night,
>
> Is in a noyous cloud enueloped,
>
> Where she may find the substaunce thin and light,
>
> Breakes forth her silver beames, and her bright hed
>
> Discouvers the world discomfited … (*Faerie Queene* (1590) 3.1.43).

Like a comet, Cynthia appears shrouded as she arrives to observe her surroundings, but if it is as bright as the comet of 1577, it shines so brightly that it can "far outcast" its own shape, lighting the ground and casting a shadow.

In 1579, the December portion of Spenser's *Shepheardes Calender* followed this initial string of five bright comets. As part of a reflection on life and death that defines a relationship between man and nature, Spenser invokes the appearance of a comet that "stird up that unkindly heate" of both natural meteorology and human love (MacCaffrey 554–556). But even Spenser's words are presaged by the words of Fleming, who delicately blended comet literature with science.

Comets and Thomas Nashe

"In all points our brains are like the firmament," Thomas Nashe records in *The Terrors of the Night (1594)* "and exhale in every respect the like gross mistempered vapours and meteors, of the more foeculent combustible airy matter thereof…." In Nashe's simile, the human mind is a multifarious free-flowing amusement park where ideas and images are born, fly about, and either disappear or evolve into other images. Nashe's comparison of such complexity with the atmosphere of the Earth is original and insightful, even by today's standards. To this cauldron he adds comets and meteors, which we have already seen were considered, despite Tycho's conclusions from the comet of 1577, to be Aristotelian creations of the atmosphere. One could suspect that by vapours, Nashe is referring to Aristotle's theory that comets can be thus explained, as paraphrased from *Meteorologia 7*: The dry, warm exhalation of air is the outermost part of the terrestrial world which falls below the circular motion. In the course of this motion it often ignites, and this he maintained was the cause of the 'shooting' of scattered 'stars.' In the very next sentence, Aristotle continues: if the upper motion introduces a kind of fiery principle not so strong as to burn up the material quickly, nor so weak as soon to be extinguished, then a comet is formed. Aristotle's explanation seems to fit well with Nashe's phraseology of "mistempered vapours (for comets) and meteors." Nashe is using Aristotle's explanations as a scientific basis to begin his criticism of the soothsayer who would imagine our brains creating "monstrous images" of the night.

Nashe's *Anatomy of Absurdity* was published in 1588 (Steane, 824–828), shortly after the appearance in October 1587 of a bright comet with a tail that stretched over more than a quarter of the sky (Yeomans 416). Nashe seems to criticize those who study Nature, but the examples of Nature that permeate his writings show at least that he is more than casually aware of natural events, especially earthquake and celestial event phenomena. As an example, Nashe captures a dream of many comets: "He falleth asleep; no star he seeth in the night but seemeth a Comet; he lighteth no sooner on a quagmire, but he thinketh this is the foretold earthquake, whereof his boy hath the ballad." Nashe alludes to the ideas of one of the earliest astronomers, Thales Milesius (b. circa 624 BCE): "they see not what is under their feet, searching more curiously into the secrets of nature, whenas, in respect of deeper knowledge, they seem mere naturals." Perhaps this can be taken as a prelude to the discussion of the predictive value of eclipses in *Lear* (see Chap. 3). Thales the judicial astrologer is criticized, not Thales the great natural astrologer and careful and thorough observer of the night sky.

Three centuries later, the American astronomer Barnard, in the midst of the apparition of the great comet of 1882, would experience a similar dream. When Barnard went to sleep early in the morning of 14 October, he dreamt about seeing many comets. When he awoke some hours later he turned his telescope to the great comet, then moving his telescope to the southwest and discovered a group of a half-dozen small comets accompanying the great one. Barnard wondered whether he had fallen asleep on his feet and had resumed his dream. The comets were genuine enough, and Barnard's report was confirmed by observers in Europe. But although the little comets moved at the same rate and direction as the great comet, they all faded within a day and eventually vanished. The Barnard dream harks back to Nashe in the sense that both writers imagined a celestial stage filled with comets.

In 1594 Nashe's *Terrors of the Night* (530–531) includes references to "blazing Comets" and the "fiery streaks" of meteors. Nashe's frequent references to meteors are included in this chapter because of the old Aristotelian idea that they, like comets, belong to our atmosphere. At the end of the sixteenth century the relation between the two was based on that principle. Four hundred years later, we understand that while meteors are engendered within our atmosphere, the dust that forms them comes from comets. These comets might well be references to the comets that appeared in 1592 and 1593 (Yeomans 416). In *Summer's Last Will and Testament* (circa 1592) Nashe offers a look at inevitability of death, especially in his plague-ridden time. Autumn asks "Who treadeth not on stars, when they are fallen?" (*Summer's Last Will*, 198) and the "rough and stern" Winter's son Backwinter wishes that

> I could bark the sun out of the sky,
>
> Turn moon and stars to frozen meteors
>
> And make the ocean a dry land of ice; … (*Summer's Last Will*, 201)

Nashe's allusions to meteors in *Summer's Last Will* comes just 2 or 3 years before Shakespeare's *Richard II* (1595). In successive lines, the Captain invokes meteors and a lunar eclipse to assert the judicial power of the cosmos (2.4.7–11):

> Captain: 'Tis thought the king is dead; we will not stay.
>
> The bay-trees in our country are all wither'd
>
> And meteors fright the fixed stars of heaven;
>
> The pale-faced moon looks bloody on the earth
>
> And lean-look'd prophets whisper fearful change …

A few years later, *Hamlet's* Horatio recalls a scene from the earlier *Julius Caesar*, then adds further lore to it (*Hamlet* 1.1.114–120):

> … A little ere the mightiest Julius fell,
>
> The graves stood tenantless and the sheeted dead
>
> Did squeak and gibber in the Roman streets;
>
> As stars with trains of fire and dews of blood …

Either comets or meteors could have "trains of fire," though the reference functions better if the image conjures up a rushing meteor rather than a comet parading slowly and majestically. The sheeted dead moving aimlessly through the streets of Rome is a better match for meteors falling from the sky.

All these references were too early to have been inspired by the great Leonid meteor storm of November 1602 (Kronk, Comets and Meteor Showers), though Shakespeare's meteor images "fight the fixed stars of heaven" and "stars with trains of fire" are so vivid that I suspect they emanate from his personal observation. There are few contemporary English references to sixteenth century meteors. If a major shower of meteors did take place or perhaps a series of a few bright meteors or fireballs in the early 1590s, it is supported only by an increased number of references to meteors by Nashe and Shakespeare.

The idea that comets and meteors are physically related was given substance by the discovery of the periodic Comet Swift-Tuttle, which was discovered by Kegler in the eighteenth century and by Swift and Tuttle in 1862, and whose orbit closely matched those of the Perseid meteors which add eloquently to the beauty of the August night sky each year. Three centuries earlier, that connection was understood in a different way, as John Bainbridge tells us in his *Astronomicall Description of the late Comet from the 18. of Novemb. 1618 to the 16. of December following*:

> … let this Comets regular and ordinate motion (for a month together)
>
> never deviating from one right line, keepe you in the way of verity. This very reason alone many ages agoe persuaded divine *Seneca*, and in our Fathers daies that ingenious and subtle *Cardane* to place all Comets above the elementary regions, wherein only inconstant and momentany (sic) Meteors make their fickle vagaries.

In Bainbridge's world, comets and meteors are related in that both are meteorological, with comets inhabiting the atmosphere's upper regions and the inconsequential meteors decorating the air closer to the ground.

Nashe's friend George Peele, whose works Nashe praised as *primus verborum artifex*, lets comets brighten up his *Araygnement of Paris* (1585). By comparing the plant life that grows along the banks of "bubbling brooks" like water-flowers and lilies, to the mighty comets, he invokes a unity in Nature that brings together the infinitesimal and the infinite. The lines compare the tiniest of objects that requires a keen eye for detail to spot, to one of the great comets that slid across the sky between 1573 and the poem's probable date of composition of 1585. At the start of the passage Iris is invoked—"Iris adornes her arch"—i.e. the arch of the heavens, perhaps with "pride and bravery." The comparison begins with the cosmic—"milk—white way" of course the Milky Way that

in frosty night Appear so fair and beautiful in sight,
which then is focused inward to these fields, and groves, and sweetest bowers
Bestrew'd and decked with parti-colored flowers.

Now that the poet has us focusing inward, he suggests motion: The flowers are not just appearing, but gliding and sliding.

> Along the bubbling brooks and silver glide,
>
> That at the bottom do in silence slide;

Watching this motion are

The water-flowers and lilies on the banks …

The poet thence takes us outward again. Small as they are, the water-flowers and lilies shine

like blazing comets, burgeen all in ranks;

Under the hawthorn and the poplar-tree

Where sacred Phoebe may delight to be … (Peele (1585))

The microcosmic scene, with flowers standing guard next to a flowing stream, contrasts with the macrocosm, where blazing comets move across the sky as they stand guard next to the fixed stars. The movement from micro to macro is instantaneous; in a single word, "shine," the focus shifts to the vastnss of the heavens. Although the simile is reversed—it is the comets that move in the sky, but the flowers to which they are compared do not move. The comparison is beautiful and accurate, apt, and precise in its details. Note also that Peele refers to comets, in the plural.

Comets After 1602

A search for comet references in literature reveals as much about the nature of comets by what is not discovered as by what is found. My search for comet references in literature shows a decline in the first years of the seventeenth century compared to the prior decade. Comets appear less often in the sky, and are harder to find in literature, after 1602. After a 5 year hiatus, the sky was altered again by the visitation of a great periodic comet which would, two revolutions later, be named for Edmond Halley, the man who calculated its orbit and discovered its periodicity. The next year Shakespeare probably wrote *Pericles*, which offers a tribute:

My lord, that ne'er before invited eyes,
But have been gaz'd on like a comet. (5.1.84–85)

Pericles functions indirectly as a telescope through which we can look back at a sky graced with an unusually high frequency of comet appearances. We read in Charles Fitzgeoffrey's *Sir Francis Drake His honorable Life*,

Where he a new-made star eternallie
Shall shine, transparent to spectatours eie:
A fearefull comet in the sight of Spaine,
But shall to us a radiant light remaine. (15.11–14)

Drake, who died in 1596, is remembered for his writing on both the supernova of 1572 and the "fearfull comet" whose appearance over Spain presaged defeat for their Armada. From the viewpoint of observation, the comet seen in 1499, but possibly not again for five centuries, can hearken us back to that earlier time. Scientists record the apparitions of these comets, but poets and playwrights bring them to life (Fig. 2.1).

Chapter 3
These Late Eclipses

Hung be the heavens with black, yield day to night! (I Henry VI.1.1.1)

In building a picture of a magnificent sky, a "majestical roof fretted with golden fire", (*Ham.*2.2.302–303) Shakespeare's opening lines suggest the wonders both of eclipses, and, in the following lines, comets. The sky was a busy place in the seventeenth century's opening years, but of the string of celestial events, the passing of great shadows closest to the Earth that we call eclipses, generated the most interest. Eclipses of the Sun and Moon were (and are) specific events that could be predicted and traced to particular dates. This opening line of the *Henry VI* trilogy, cited above, invites Shakespeare's audiences to conjure up a sky. I do not believe that Shakespeare intended the transition in the first line, from day to night, to be the gradual one that we see at day's end. Besides the obvious connection to the design of his theatre, where a black curtain literally hung from the ceiling and could be thrust into position (Greenblatt *et al.* 86), in Shakespeare's art, day hurtles into night, typical of his ideal or heroic nature passages where his focus is on his characters, not his setting; "it is as though the poet-dramatist goes to the extent of personifying natural phenomena when he dwells at any length upon them" (Spencer 43–46).

I offer two examples of heroic word paintings; the first comes from Edward III, a play with which Shakespeare was but marginally involved. If we accept the idea that he did compose scene 13 (Oxford, 257), then we are treated to a magnificent Shakespearean allusion to the appearance of a landscape during a deep partial solar eclipse:

> *King of France: A sudden darkness hath defaced the sky, The winds are crept into their caves for fear, The leaves move not, the world is hushed and still, The birds cease singing and the wand'ring brooks Murmer no wonted greeting to their shores. Silence attends some wonder and expecteth That heaven should pronounce some prophecy. Where or from whom proceeds this silence, Charles?*

D.H. Levy, *The Starlight Night*, Astrophysics and Space Science Library 419,
DOI 10.1007/978-3-319-19878-1_3

> *Dauphin: Our men with open mouths and staring eyes Look on each other as they did attend Each other's words, and yet no creature speaks, A tongue-tied fear hath made a midnight hour, And speeches sleep through all the waking regions.*
>
> *King of France: But now the pompous sun in all his pride Looked through his golden coach upon the world, And, on a sudden, hath he hid himself, That now the under earth is as a grave, Dark, deadly, silent and uncomfortable.* (*Edw. III*.13.1–18)

This allusion comes close to describing a scene during a partial eclipse of the Sun, or even a total eclipse just before the onset of the total phase. The "sudden darkness" thrusts itself more suddenly than a thick cloud deck does before a thunderstorm, and I have seen the wind drop precipitously so that the world, indeed, "is hushed and still," and birds literally "cease singing." These changes appear suddenly as the crescent Sun continues to thin.

However, this play has been dated at 1594, which is almost two decades before the deep partial eclipse of 1605. It is not likely that Shakespeare had witnessed such an eclipse before he wrote these lines. I suggest that he based the analogy either on his wide reading (Holinshed's *Chronicles* document at least four such eclipses), or perhaps that some traveler reported his experiences to him of the brief total eclipse on the Atlantic during the afternoon of 29 April, 1585, or the longer total that crossed the south Atlantic Ocean on 30 May, 1593 in the path of English sailing ships (Espenak, <http://eclipse.gsfc.nasa.gov/SEcat5/SE1501-1600.html>).

The opening of *I Henry VI* is a second example of a heroic word painting. The sky descends to black as though a celestial curtain, in this case the shadow of the Moon, has suddenly been thrust upon it. It could be interpreted as an ideal or emblematic image of the sudden darkening of the sky due to an eclipse that accompanied the death of Henry V.

Not many years later, Shakespeare summons this image again, this time in a slide to the darkness of eclipses in the second scene's opening line in *King Lear*: "Thou, Nature, art my goddess," Edmund declares, "to thy law/ My services are bound" (1.2.1–2). As Gloucester complains of the deteriorating situation in Lear's palace, he cites Kent's banishment as his instigation, then goes on to France's anguished departure and the King's unpredictable behavior, and finally demands to read Edgar's letter from Edmund's hand. As he scans it, Gloucester uncovers Edgar's suave delight in finding a perverse satisfaction in "aged tyranny." Edmund probably forged the letter himself, blaming its instigation on his brother and thereby leading to Gloucester's damning his son as a villain. Edmund then conveniently offers to intervene, to learn about Edgar and report back to the Earl.

Gloucester then undergoes a profound change. From anger towards Edgar, as he peers across the stage his mind begins to focus on the consequences of the recent eclipses. He has almost reached a state of panic as he intones: "These late eclipses in the Sun and Moon portend no good to us. Though the wisdom of nature can reason it thus and thus, yet nature finds itself scourged by the sequent effects: Love cools, friendship falls off, brothers divide" (1.2.101)—all thoughts that refer to obvious motifs from the play that include Lear's rage at Cordelia and Kent, and the developing rift between Edmund and Edgar. After Gloucester exits, Edmund dismisses his father's reasoning, countering the older man's "portent" with his own "conceit": "This is the excellent foppery of the world, that when we are sick in fortune, often the surfeits of our own behavior, we make guilty of our disasters the sun, the moon, and stars, as if we were villains on necessity" (1.2.115–127).

Edmund's skeptical-rationalist perspective solidifies the eclipse theme that reached an audience that had the recent lunar and solar eclipses in mind, and suggests that astral influence is not at work during this play. The discussion reads like a sixteenth century precursor of a modern radio talk program; Edmund's speech characterizes the English people who believe in the predictive power of eclipses as "fools by heavenly compulsion, knaves, drunkards, and treachers (sic) by spherical predominance." In those two adjacent speeches, Shakespeare offers us ideas about the nature of eclipses and their effects on humanity. However, his conclusion is, like so many aspects of Shakespeare, left unwritten.

It could be said that the events in the play itself supported a conclusion that the eclipses did portend the disaster in the following acts, the blinding of Gloucester, and the deaths of Cordelia and of Lear. I suspect that Shakespeare left the argument hanging purposefully, considering it sufficient that the issue is raised between Gloucester and his sons in a time just before or after the eclipses in England. Shakespeare, as he should, posed many questions about solar, lunar, planetary, and astral influence, then left them unanswered. It is my belief that the playwright simply wished to open up possibilities for his audience. As a playwright, it was not Shakespeare's job to define public views on various matters, but instead to tell a story that might illustrate the fates at work.

By the early seventeenth century the mechanics of eclipses were well understood. As the Moon orbits the Earth once each month it can pass between Earth and Sun in its new phase, and at that time a solar eclipse can occur. Two weeks later, the Moon, now full, can pass into the shadow of the Earth causing a lunar eclipse. Because these eclipses repeat every 18 years, 11.3 days, one can predict accurately the occurrence of eclipses far into the future.

The rich sequence of eclipses in 1605 began during the evening of 13 April (OS) with a long total eclipse of the Moon. There was a second lunar eclipse, a partial one, during the morning of 27 September (17 OS); finally there was a deep partial solar eclipse on 12 October (2 O.S.). That event was total in a narrow path that began at sunrise over Baffin Island in the north Atlantic, crossed over southern France then through the area around Barcelona, Spain, Nicosia in Cyprus, and left the earth at sunset over the western shore of the Caspian Sea, all within about 2 h.

King Lear invokes at least the final pair of a lunar and a solar eclipse, or possibly all three events, and in Sonnet 35 Shakespeare notes how "Clouds and eclipses stain both moon and sun," where the order of the autumn 1605 pair of eclipses is correct. The uncertain date of Sonnet 35 casts doubt on which eclipse he might have been referring to, though we do know from John Dee (61) that the solar eclipse of 1599 was completely obscured by clouds, and from Pingré (37) that the end of the 1605 solar eclipse, at least, was observed from London.

The October 1605 eclipse was not total in London, but it was close enough that the following effects should have been observed: as mid-eclipse approached the Sun would have appeared as a rapidly thinning crescent, the location of the crescent rounding the center of the Sun, like a minute hand on a giant celestial clock in a clockwise direction from east to west. As it thinned, the sky darkened rapidly and noticeably, indeed as if "dark night strangles the traveling lamp" (*Macb*.2.4.10). The eclipse did not have to be total for Shakespeare's powerful imagery to work. In the absence of a total eclipse, dark night would strangle, then as quickly withdraw.

Despite the fact that the solar eclipse of October 1605 was not the strongest or longest of this series, it attracted more attention by far than any other eclipse during that period. In early October the sky over London was likely to be less cloudy than at other times during the year, allowing a greater number of people to see the eclipse.

Dade's 1605 almanac predicts malicious but unspecific effects from this eclipse that might be contradicted or nullified by appropriate lifestyle changes: "I omit to speake of, onely beseeching God to direct us with his holy spirit, that we may live as true Christians, faithfully processing the Gospell of Jesus Christ ..." (Dade B3 +7). Gloucester's speech goes further, delineating the kinds of disasters to which an eclipse can lead, beginning with the same strife between parents and children that Lear is feeling. In February 1606, Edward Gresham told of strange events in Croatia that included a report of a woman giving birth to a boy who had four heads, a "confirmation" of the events that followed the eclipse series that ended four months earlier. The extent to which a belief in *astrologia judicialis* influenced these observations will be explored in Chap. 4.

The eclipses in *King Lear* offer an entrance to studying *King Lear* from a New Historicist's point of view. We know that new historicism connects a work of literature to the culture in which it was written, and by inference its politics and chambers of power; in the time of *King Lear*, judicial astrology was a major part of that power. Shakespeare wrote *King Lear* at a time when judicial astrology was deeply frowned upon, but also at a time when astrology was flourishing among England's general population. Throughout Europe, astrologers were a fixture in courts that were often decorated with symbols portraying the Zodiac. Greenblatt developed his new historicism ideas to reflect his own "tendency ... away from a criticism centered on 'verbal icons' toward a criticism centered on cultural artifacts." (Greenblatt 3) Greenblatt is vague as to what defines a cultural artifact; the idea of astrology is one, and the Moon, Sun, and their eclipses taking place in his time are another. In his speech, Gloucester offers good examples of how these eclipses, and their "sequent effects" can change our lives. Love cools as eclipse-associated darkness diminishes the intensity of interpersonal relationships, and treason arises in palace courts, especially that of James I, whose 1605 parliament is threatened to within an inch of its life by the gunpowder plot of 5 November. Even though the threat was deadly serious, James's letter from a month earlier was lighthearted, even teasing, in its tone, and he admitted that his own sense of humor was affecting his ideas that day, as he wrote:

> But now will I go to higher matters, and tell you what I have observed anent the effects of this late eclipse, for as the troglodytes of the Nile that dwelt in caverns, the shepherds of

> Arcadia dwelling in little cabins, the Tartars harbouring in their tents like the old patriarchs, so I, having now remained a while in this hunting cottage, am abler to judge of astronomical motions than yew that lives in the delicious courts of princes. The effects then of this eclipse for this year are very many and wondrous. It shall make divers noblemen at the Court loathe their wives and wish they were better married, such as Lennox, Pembroke, and Roxburghe. It shall make some widowers loath to marry again. (Akrigg, 265)

The letter ends with his wish to return to London and get on with the opening of parliament on 5 November. Considering that he barely escaped that opening with his life in the wake of the gunpowder plot, he might have preferred staying at his cottage.

Gloucester relies on the judicial astrology of the time to buttress his argument about malevolent eclipses. Even now, astrologers tend to view eclipses as large-scale signals of coming danger. Some astrologers followed a narrow application of their craft to eclipses, one blaming a single undated lunar eclipse for the devastation of all the land of Chaldea (MacNeice 110). Later believers of Gloucester's view would cite two solar eclipses in 1781, an annular eclipse in April and a total in October, as leading to the critical environmental damage to islands off the western coast of India.

Edmund's disagreement with his father was as unmitigated as a total eclipse. His speech confirms Danby's inferring Shakespeare's interest in Nature in light of the words spoken by characters such as Edmund. By ascribing our difficulties to the eclipses, we deny ourselves the opportunity to atone for sins which are far more often "the surfeits of our own behavior" and mistaken actions. This is not evidence that Gloucester's illegitimate son has any sympathy for the misadventures that have already occurred, such as Kent's being banished, or for the more disastrous ones to come. Returning, Edgar inquires what his brother has been up to, and Edmund admits: "I am thinking, brother, of a prediction I read this other day what should follow these eclipses." With some surprise Edgar asks, "Do you busy yourself with that?" Yes, the answer comes both from Gloucester and Edgar, but in opposite ways; Gloucester's shock over Edmund's independent view, and Edgar's feigned brotherly concern that Edmund will interfere with his plan. Another philosopher of the time, Francis Bacon (112), notes peripherally how "the astronomer hath his predictions … of eclipses" in a phrase echoed in *Faerie Queene*:

> But true it is, that when the oyle is spent
>
> The light goes out, and wicke is throwne away … (*FQ* 2.10.30)

Some critics, notably Arden, suggest that the eclipses can be used to date the play more accurately than between 1603, when Samuel Harsnett's *Declaration of Egregious Popish Imposture* appeared, and Christmas 1606, when the play was performed before King James (Greenblatt, 2326). In his biography *Shakespeare Unbound*, René Weis cites the evidence of the eclipses as positive dating of the play at late 1605. "Gloucester is unmistakably referring to those" in his "late eclipses" speech; "the play can be dated to after the eclipses …" (Weis 334). Although I guess that Weis is right, I think that his evidence is skimpy; Shakespeare could just as well have been consulting an almanac by Dade or someone else, and that would date the play just as easily *before* the eclipses.

Other Contemporary References to the Eclipses of 1605

My search of contemporary references to the eclipses of 1605 found few surviving examples other than those in Holinshed and Shakespeare. The three eclipses were announced in Dade's *Almanac* (n.p.) which was widely available by the early part of 1605. The spring total eclipse was trumpeted as an "Eclipse of the moone appearing the 24 day of March at 7 of the clocke at night …" The September eclipse was "appearing, and being seene unto us, above our horizon the 17 day of September at 3 of the clock, 37 minutes in the morning, She being darkened almost 9 points, and from the beginning of her darknesse into the recoverie of her former light, will be about one houre and a halfe …. The influence of this Eclipse will begin to worke and take effect the 17 day of August the next following, and shall continue until the 5 day of October then next ensuing."

The October solar eclipse appears thus: "The Sun eclipsed the second day of October, at one of the clocke 20 minutes in the afternoone, whose obscuritie and darkness, will be 9 poynts and 10 whose obscuritie and dark beginning to the end one hour… ." (Dade n.p.)

There has been commentary on the idea that these two eclipses could have been, but were not necessarily, the events Shakespeare had in mind when he wrote *King Lear* (Parr and Harbage 1060, Foakes 91, Kermode 1250). Johnstone Parr and others suggest that all of the three solar eclipses that occurred around the writing of *Lear*—in 1598, 1601, and 1605—were the events Shakespeare had in mind. These commentaries typically end with the conclusion that since eclipses were not uncommon, Shakespeare's reference could have been to the phenomenon in general, not to any specific eclipses, or that Shakespeare intended a general reference to the sudden occurrence of several eclipses after a half-century lull.

I disagree. Just as it is unlikely that a play written today would make much of the great European solar eclipse of 1999, I doubt that in 1604 or 1605 Shakespeare would have invoked his audience's memory of eclipses that had occurred in 1601 and 1599, especially since John Dee tells us that the 1599 eclipse was clouded out. The reason I believe that the eclipses of the Sun and Moon in 1601 were not a factor is that it has been established that both Harsnett's *Declaration of Egregious Popish Impostures* and Montaigne's *Essays*, translated by Florio, had an effect on *King Lear's* language (Foakes 91). These books were both published in 1603, which would have made the specific remarks about eclipses of both the Sun and Moon appear dated even if the play had been written and performed during that year (Furness 186, 377).

There are difficulties in trying to use specific astronomical references, especially those about eclipses, to date Shakespeare's writings. For example, in his 1938 article in *Shakespeare Quarterly*, Alfred Harbage proposes that Sonnet 107's "mortal Moon" alludes not to an eclipse at all but to the crescent-shaped battle formation of the Spanish Armada, and then cites earlier sources claiming that the allusion is to the Armada's defeat in 1588. Citing that the Moon was the symbol of Elizabeth (Cynthia's imperial vot'ress), Harbage proposes that the "mortal Moon" actually refers to Elizabeth. (Harbage 59) He goes on to cite how other factors in the sonnet encourage a much later date, perhaps including Queen Elizabeth's death in 1603. I agree with this interpretation, particularly since an actual lunar eclipse was visible from England on 24 May 1603 (Espenak).

Other Eclipses

During the predawn hours of 10 February 1598 (O.S.), a total eclipse darkened the Moon over England. Only 2 weeks later, an almost total eclipse of the Sun passed near England. John Dee's marginal notes detail his attempt to see that event:

"Feb. 25th, the eclips. A clowdy day, but great darkness abowt 9½ mane" (Dee, 61). The darkness was understandable; had the sky been clear Dee would have observed a crescent Sun so thin that only a tiny sliver of sunlight could be detected behind the dark sphere of the Moon. On 10 July 1600, London experienced a noon hour partial solar eclipse during which almost half the Sun was obscured from the Moon. Eighteen months later, on 24 December 1601. London was near the path of an "annular" solar eclipse during which the Sun would have appeared as a sickle of light, almost a complete ring of light in the sky. I have found no reports about this event, but I did note writing about stormy weather at the time, indicating that the eclipse was probably completely clouded out. The almost total lunar eclipse of 1601 was one of the darkest on record. In the early evening hours of December that year, almost the entire Moon was enmeshed in the Earth's shadow, and the dark portion simply disappeared.

Written around 1600, *Hamlet*'s opening lines are replete with night sky imagery. The star "westward of the pole" has been discussed in Chap. 1, but later in the same scene Horatio invokes the memory of those events that could accompany the visitations of ghosts:

> stars with trains of fire, and dews of blood,
>
> Disasters in the sun, and the moist star
>
> Upon whose influence Neptune's empire stands
>
> Was sick almost to doomsday with eclipse. (*Hamlet* 1.1.117)

Stars with trains of fire are more likely interpreted as meteors than comets, and the moist star, or the Moon, is "sick … with eclipse" as it was in 1598 and again in 1600. Written circa 1603 or 1604, *Othello* is rich with references mostly to lunar eclipses. There is a good chronological reason for this: Between 1600 and 1605, no solar eclipse reached England, but a partial lunar eclipse took place in the late evening hours of 14 May 1603 (O.S.). Just past 11 in the evening the Moon faded dramatically, and over half of it was a ruddy red color. Six months later, an early evening November partial eclipse darkened a small portion of the Moon over England.

Othello's references to eclipses begins with the protagonist's plan, "To follow still the changes of the moon with fresh suspicions" (3.2.130–133). I suspect that combined with the other references to eclipses that occur through this play, the "changes" do not allude merely to the ordinary change of lunar phase but to eclipses. This suspicion is strengthened in Act 4, when he watches heaven as it "stops the nose at it, and the moon winks … (4.2.76), fading and then brightening again as during an eclipse."

The almost total eclipse of the Moon on 9 December 1561 was listed as one of darkest eclipses on record. Visible from London the evening of December 9 (*Sky & Telescope* 142, March 1964, 142–146), the eclipse became an "error of the moon (that)comes more nearer earth than she was wont And makes men mad." (5.2.112)

And as Othello nears death, he

thinks it should be now a huge eclipse

Of sun and moon, and that th' affrighted globe

Did yawn at alteration. (5.2.101–103)

Eclipses come in 18-year cycles (Greek *saros*) that last hundreds of years. The saros that included the 1605 eclipse repeated itself 18 years, 11 1/3 days later in 1623 and at equal time intervals after that. In 2002 I witnessed the 22nd repetition of that 1605 solar eclipse. What I saw was a much shallower partial eclipse than the one visible over southern England four centuries earlier. The sky still darkened significantly so that during the central portion of the eclipse the Sun's light was considerably dimmed, shadows were sharper than usual, and the landscape adopted a strange unearthly quality of light. The preceding lunar eclipse occurred on 26 May 2002 followed by the solar eclipse on 10 June 2002. Because of the Earth's rotation during the additional third-of-a-day, however, neither of these eclipses was visible from England.

Londoners would have seen an even more dramatic spectacle if the sky were clear on that 1605 afternoon, when more than 90 % of the Sun was covered by the Moon. Within just 5 years, eclipses would take on a different meaning for those who had the means to view an eclipse through a telescope. The deep partial eclipse during the morning of 30 May 1612 provided an opportunity for Londoners to revisit *King Lear* with the aid of a telescope.

Shakespeare's most eloquent description of an eclipse does not even mention the word (*Macbeth* 2.4.20). An eclipse seemed to be associated with the word "disaster," which has astrological connotations of "bad star" and occurs elsewhere in Shakespeare. (e.g. *Macbeth* 3.1.110–112) Shakespeare apparently reasoned that by taking advantage of contemporary astronomical events, he could engage his audiences' interest in the narratives; by the fall of 1605 the English public would not have been shocked by the eclipses, though perhaps they took note that there were so many of them, particularly after a dearth of eclipses that had stretched over much of the previous half century. If *Macbeth* was indeed composed in 1606, after the 1605 solar eclipse, this description would be consistent with the idea that Shakespeare might have seen a near-total eclipse in October 1605, as seen in the thane of Ross's speech after Duncan's murder:

By the clock, 'tis day,
And yet dark night strangles the traveling lamp:
Is't night's predominance, or the day's shame,
That darkness does the face of earth entomb,
When living light should kiss it? (*Macbeth* 2.4.6–10)

How exactly does dark night strangle? A dark cloud or approaching storm would not "strangle" the Sun as much as it would generally and more gradually darken the sky. Shakespeare's heroic imagery calls up a vision that explicitly affects the Sun by suddenly cutting off its light. In the half hour preceding and following the maximum phase of the eclipse on 2 October 1605, the onset of a twilight glow unlike that of any approaching storm would easily fit the conditions to which Shakespeare alludes. The dark night does not descend quickly *because* of a natural fall from sunset or from a rapidly approaching storm; it follows in an instant from the Moon's passing in front of the Sun. Discussion about the meaning of these eclipses begins with astrology: Moberly was still grieving over the loss of Queen Elizabeth I, who died in 1603. If we are to believe any of the critics of that time, then we can understand how the loss of the Sun, even for a short time, was seen as catastrophic. The allusion to the darkness of an eclipse can be carried one step further. When Macbeth learns of his wife's death he utters one of the most famous and brilliant speeches Shakespeare ever wrote:

She should have died hereafter,
There would have been a time for such a word.
Tomorrow, and tomorrow, and tomorrow
Creeps in its petty pace from day to day
To the last syllable of recorded time
And then is heard no more. Out, out, brief candle.
Life's but a walking shadow, a poor player
That struts and frets his hour upon the stage,
And then is heard no more. It is a tale
Told by an idiot, full of sound and fury,
Signifying nothing. (5.5.22–27)

This speech can be interpreted in relation to Ross's earlier notion about a solar eclipse. Macbeth magnifies the allusion; the shadow envelops not just a portion of Earth but all human life. Instead of moving swiftly and efficiently the shadow, as life, now "struts and frets." As A. D. Nuttall points out, "While the great mystics speak of a world suddenly enhanced, blazing with fresh significance, Macbeth describes the draining away of all meaning from the universe" (Nuttall 287–288). This eclipse allusion, metamorphosed from the shadow of the Moon to a definition of human life, is called on once again as a metaphor to mirror the cheapness of a life that has lost all meaning. *Antony and Cleopatra*, also written around 1606, shows how

> Our terrene moon
>
> Is now eclips'd, and it portends alone
>
> The fall of Antony. (Ant. 3.13.154)

A Treasure of Eclipses in Holinshed's Chronicles

Virtually every high school student of Shakespeare knows that Raphael Holinshed's *Chronicles* was one of the Bard's most important sources. Published in 1577, the three-volume work is a detailed account of the history of England, Scotland, and Ireland. Eclipses are mentioned frequently within its pages. Although no eclipses are mentioned in Holinshed's account of Macbeth, who ruled from 1040 to 1057, (Holinshed II, 168ff) there was a three-minute-long total eclipse in Scotland on 24 January 1023. (There is a "darke night" in Holinshed (II, 172) but it commemorates not an eclipse but the night of Banquo's murder.)

The "black" curtain in the first line of *I Henry VI* harks back to Holinshed's "black hour" in his account of the year 1433 of *The Historie of Scotland*. "In the same yeere the seventeenth day of June, was a terrible eclipse of the Sunne, at three of the clocke at after noone, the day being darkened over head for the space of one halfe houre together, as though it had been night, and thereupon it was called the blacke hour." There was indeed an eclipse over Scotland on 17 June 1433. The eclipse had a long total phase lasting 4½ minutes and traversed the entire northern half of Great Britain. Can an eclipse yield day to night that quickly? In all eleven total eclipses of the Sun I have witnessed, day yielded to night as quickly as if someone were adjusting a dimmer-style light switch. When Shakespeare wrote these lines around 1591, however, he had not seen a total eclipse. Although it is possible he glimpsed, in his youth, a partial eclipse at sunset on 29 April 1585, it would not have been hard for him to have read about one. Interest in eclipses was substantial in Shakespeare's time, both as natural events and as portents. Anyone interested in historical eclipses of the Sun and the Moon that were successfully viewed from Britain could have found them in the *Chronicles*, which offered several easily accessible accounts of eclipses dating back to the 1100s. I have reconciled Holinshed's dates and times with those of NASA's eclipse site (http://sunearth.gsfc.nasa.gov/eclipse/eclipse.html), and most of them are accurate to the day and hour. Reading the ancient *Chronicles* provides a human perspective to the NASA data and maps of these events.

Shakespeare and his contemporaries did not have access to NASA, but they certainly could have studied the eclipse stories in Holinshed as easily as we can. Each volume has an index giving the location of information (like eclipses) to within a few lines on a page; one enthusiastic sixteenth century typesetter even substituted a sketch of the 1430 partially eclipsed Sun for the letter "c" in the index entry for the "eclipse of the Sunne terrible." (II, index n.p.). One further reads that on 14 May

1230, "marvelous eclipse of the Sunne … immediatelie after the rising thereof so that the earth seemeth as it had beene covered againe with darke of night. On that morning dawn came, then pushed itself back as the eclipse deepened until the Sun rose in total eclipse." The NASA website, prepared four centuries later, confirms that the shadow of the Moon dropped from the sky at sunrise over Scotland on that memorable day. Later that year, "On the 22nd daie that November, the Moon was … eclipsed" (Holinshed 3, 212). Holinshed's detailing of eclipses ends with the eclipses of 1544: "This yeare chanced four eclipses, one of the Sunne the fourteenth of January, and three of the Moone." From London, the solar eclipse happened on the 24th, not the 14th (O.S.) of January, and was nearly total. It was preceded by an hour-long total eclipse of the Moon, and followed by another total lunar eclipse at moonrise on July 4, and yet a third total lunar eclipse before dawn on 29 December.

Besides its rank as likely the richest source of information about some great eclipses in English history, the *Chronicles* also cites several examples of eclipses being associated with other unusual phenomena like a storm or earthquake. The "black houre" eclipse already mentioned was over Scotland. In its third volume, the *Chronicles* gives Shakespeare access to three sources for ancient eclipse information that were followed by other natural events. During the time of Henry I, on 2 August 1133, there befell

> a wonderfull and extraordinarie eclipse of the Sunne and Moone appeared, in so much that Wil. Malmef, who then lived, writeth that he saw the starres plainlie about the Sunne at the verie time of the eclipse. On the Friday after such an earthquake also happened in this realme, that manie houses and buildings were overthrowne. This earthquake was so sensible; or rather so visible, that the wall of the house wherein the king then sat, was lift up … & at the third it setled it selfe againe in his true place. Moreover at the verie same time also fire burst out at certaine riffes of the earth, in so large flames, that neither by water nor otherwise it could be quenched (III, 44).

Additionally, there was a partial solar eclipse on 13 September 1178 that was total in southern France. In England the bodie thereof appeared as it were horned, shooting the horns to the west as the moone doth … the horns at length were turned toward the west, and so the blacknesse awaie, the Sunne received his brightness againe. In the meantime, the air being full of clouds of diverse colours, as red, yellow, greene, and pale, holpe (sic) the people's fight with more ease to discerne the manner of it. (That eclipse is part of a Saros cycle, 121, which repeated as an annular eclipse over Antarctica in 2008.)

That historical event was preceded by a unique vision which involved the Moon the previous June, when the new moon shone forth very faire with his hornes towards the east, streightwais the upper horne was divided into two, out of the mids of which division a burning brand sprang up, casting from it a farre off coles and sparks, as it had beene of fire. The bodie of the (Moon) in the meanetime that was beneath, seemed to wrest and writh in resemblance to an adder or snake that had beene beaten, and anon after it came to its old state againe (III, 102).

(This story appears also in *Gervasii Cantuariensis Opera Historica: Chronica Gervasii, Rerum Britannicarum Medii Aevi Scriptores*, London, (1879) 73a.)

Geologist Jack Hartung has made the controversial suggestion that those who viewed this event had actually witnessed an impact of a comet or an asteroid on the Moon, and the formation of a new crater (Hartung 187ff).

Finally, the eclipse of 16 July, 1330 relates eclipses and storms directly, as would later appear in *Lear*. It begins with a "great eclipse of the Sunne, and for the space of two months before, and three moneths after, there fell exceeding great raine, so that through the great intemperance of weather, corne could not ripen, by reason whereof, in manie places they began not harvest til Michaelmas …" (Holinshed III, 348).

"On Christmas even, about the breake of day, a marvelous … and terrible wind came forth of the west, which overthrew houses and buildings, overturned trees by the roots, and did much hurt in diverse places" (III, 348). There was another eclipse on 21 July 1255, which was connected to Henry III's Earl of Gloucester: On 21 July 1255, the moon "suffered a marvelous eclipse … it began afore midnight, and continued foure hours. The king (Henry III) in the behalfe of his daughter the queene of Scots raised a power, and drew northwards, sending before him the Earle of Gloucester …" (III, 251).

In *Macbeth* (1606) the third witch completes a cauldron that includes a …

> gall of goat, and slips of yew,
>
> Sliver'd in the moon's eclipse (4.1.17–18)

Probably because of astrological fears about their nature, the listing and detailed descriptions of eclipses in Holinshed is an indication that natural events appeared to play a more important role in the study of history than they do today. However, Shakespeare's persistent use of them probably stems from his own innate interest in eclipses.

Shakespeare, Nashe, and the Eclipses of 1591

In his 37 plays, poetic stories, and sonnet sequence, Shakespeare invokes eclipses more often, and in greater depth, than any other writer in this investigation. Even when he uses "eclipse" as a verb, unrelated to the specific experience of an astronomical eclipse, the word invokes a sense of sadness or loss. In *3 Henry VI* written about 1591, a year with two total eclipses of the Moon, King Henry bemoans that his "joy of liberty is half eclips'd" by Margaret and Edward's failure to return from France (*3 Henry VI*.4.6.63) when King Richard, Hastings, and Stanley rescue Edward while he is hunting and depart for Flanders to seek aid. In *1 Henry VI* (written around 1591) as Talbot takes leave of his son, his metaphor is of the Sun and Moon engaging each other in eclipse:

Born to eclipse thy life this afternoon.

Come, side by side together live and die;

And soul with soul from France to heaven fly. (*I Henry VI*.4.7.53)

The eclipses of 1591 provided plenty of fodder for Thomas Nashe to use in a satirical essay. Nashe was scripting a belated response to Richard Harvey's *Astrological Discourse*, which in January 1583 predicted dire effects from the "great conjunction" of Saturn and Jupiter and which will be treated more fully in the next chapter (Nicholl, 23). The conjunction passed on 28 April 1583 without any ill effect, with the result that Harvey was embarrassed and shamed (Nicholl, 34). But beyond the conjunction, the *Discourse* also considered the effects of "the Eclipse of the Sunne, which happened the last yeare, 1582 … the bodie of the Moone at hir chaung, being directly put betweene the Sunne and the Earth, or between our sight, and the Sunne, and thereby depriving us of the full light of his beames, there appeared within our horizon at Cambridge (as you remember) a small Eclipse of the Sunne …" (*Discourse* 54). Harvey's point was not so much that there was an eclipse in 1582, but that the eclipse took place with the Sun in the constellation of Cancer. (Actually, on 20 June 1582, the Sun was not in Cancer, but in Gemini. The error in astrology was caused by a slow wobbling motion of the Earth, called precession, unknown at the time, that the Earth spins in its orbit like a slowly moving top.)

Harvey's booklet embraced judicial astrology: "there never happeneth any Eclipse of the Sunne, which doth not presignify and foreshow some great Accident to come to passe." As evidence, he cites the solar eclipse of 26 June 1424, (when the Sun also was also in Gemini) "at which time Charles the King of France being valiantly set upon in battaile by the English men and Burgundians, was expelled his owne realme. …" Harvey then mentions the more recent eclipse of 18 June 1536 (June 8 O.S.) after which there followed "at Rome, a most fearfull, and horrible thundering in the aire, being also the more extraordinarie for the very time of the yeare." He cites moreover the eclipse of 9 July 1564 (June 9 O.S.) "and effects whereof continued two years … including the Turkish invasion of Hungary and Syria." When the Sun is in certain signs, particularly Cancer, Harvey insists, eclipses are invariably followed by dire events.

Harvey's Discourse apparently remained unchallenged until 1589, when the Crown hired Thomas Nashe as a propagandist. (Nicholl, 62) Two years later, Nashe struck back with a pamphlet by one "Adam Fouleweather, Student in Ass-tronomy" a pseudonym widely believed to represent Nashe. (Nicholl, 62) In *A Wonderful, strange, and miraculous, Astrological Prognostication for this year of our Lord God, 1591*, the conjunctions were dismissed thus: "Saturne and Iupiter prov'd honester men than all the World tooke them for, whereupon the poor prognosticator was ready to run himselfe through with his Jacob's staffe" (Nicholl, see above, 34).

In his energetic rebuttal, Nashe concentrates not on conjunctions, but on the eclipses expected in 1591. "The Moon this year shall be eclipsed," the author predicts, "which shall happen in one of the 12 months and some of the four quarters of the year, whose points as they shall be totally darkened, so the effects shall be wondrous and strange." He states that the first eclipse "is little visible in our horizon" (actually the 9 January 1591 eclipse begins just as the Moon clears the eastern horizon). The solar eclipse next listed might have occurred on 25 January 1591 and which was visible only in Antarctica; "but because the Eclipse chanceth Southerly, it is little to be feared that the effects shall fall in England: yet somewhat it is to be doubted that divers Children shall be born, that when they come to age shall not know their own Fathers … , like "the bond crack'd 'twixt son and father … there's father against child." (*Lear* 1.2.98ff) As Fouleweather, Nashe points his satire against the astrology of the time "discovering such wonders to happen this year, as never chanced since Noah's flood. Wherein if there be found one lie, the Author will lose his credit for ever."" (Fouleweather, 1–7) The tract ends with a description "Of the second Eclipse of the Moon, which is like to fall out when it chanceth, either before the 31 of December or else not at all, this present year, 1591." (Fouleweather, 7)

The eclipse did take place and it turns out, was one of the most unusual in history; shortly after leaving the shadow of the Earth, the Moon also passed directly in front of Saturn. The eclipse and occultation took place on 30 December 1591 (December 20 O.S.). The rarity of this event was understood to some extent by contemporary observers. Both the eclipse and the occultation of Saturn are described in Watkins and Robertes' *A Triple Almanacke for the yeere of our Lorde God 1591* (66). "The Moone this yeere shal be Eclipsed the 20 day of December, at iii. Of the clocke in the morning, and almost a quarter, she being then in her proper motion Cancer, almost corporally conjoined with the malignant planet Saturn … which what it may presage, I leave it to others to discusse." Nashe's essay was not nearly as accurate with event dates and times as was Holinshed, but as satire it was not intended to be. As a result of the eclipse, Nashe predicted that "the Danes shall this year be greatly given to drink, insomuch that English Beer shall there be worth five pence a stoup …" (Fouleweather, 11). This was part of the "mocking prognostication" Nashe intended to extend the satire, especially as the "prophecy" about price seems a chuckling gesture to the brewers.

Even as Nashe would be branded a "restless, shifty moonlight-flitter" (Nicholl 40, 41), Nashe's satires became popular as his bitter feud with Harvey accelerated. He died in disgrace, poverty, and anonymity around 1600. Specific minor references to eclipses appear also in two of Shakespeare's sonnets:

"The mortal moon hath her eclipse endur'd" (Sonnet 107.5) could enjoin any one of several lunar eclipses Shakespeare might have seen. The other denotation, "Crooked eclipses 'gainst his glory fight" (Sonnet 60.7), is an astrological one aimed at the eclipse event in general. Eclipses do not fight, nor are they crooked, although one has the sense that time does seem to slow down, or speed up, during an eclipse. The frequency of eclipse allusions in early modern writing suggests that these sky events were not ignored, either as astrological portents or as simple celestial events. Particularly following a dearth of major solar eclipses during the second half of the sixteenth century, the celestial convergences of Sun and Moon acted to reignite interest in the sky and in literature (Figs. 3.1, 3.2, 3.3, 3.4, 3.5, 3.6, and 3.7).

Fig. 3.1 This is certainly one of the most photographed houses in the world. Known as the birthplace of William Shakespeare, it is the house in which the young William grew up. If Shakespeare did view the great new star of 1572, he would have seen it from this house or from a spot near it. Toward the west lies a modern Shakespeare research building. Photograph by the author

Fig. 3.2 Shakespeare probably went to school in this building, which still stands in Stratford not far from his boyhood home. Photograph by the author

Fig. 3.3 Shakespeare's adult home in Stratford, New Place, no longer stands, but its grounds are still delightful, in our time surrounded by a rapidly growing town. Photograph by the author

Fig. 3.4 Shakespeare's grave, pictured here, has changed little since 1616. Perhaps the light-hearted warning inscribed in the stone has helped protect it over the centuries:*Good friend, for Jesus sake forbeareTo dig the dust enclosed here.Blesste be the man that spares these stonesAnd curst be he who moves my bones.*Photographed by the author

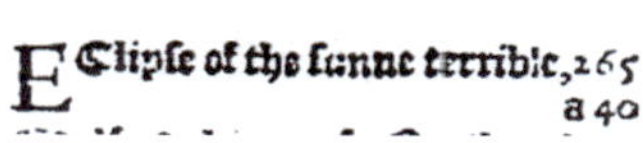

Fig. 3.5 Quite likely, Shakespeare employed Holinshed's famous *Chronicles* to learn about eclipses. Note here that the compositor assigned to the index had some fun with the "eclipse" entry, replacing the traditional lower-case letter "c" with a hand-drawn sketch of a crescent Sun. In order to know what an eclipsed Dun looked like, he probably observed an eclipse

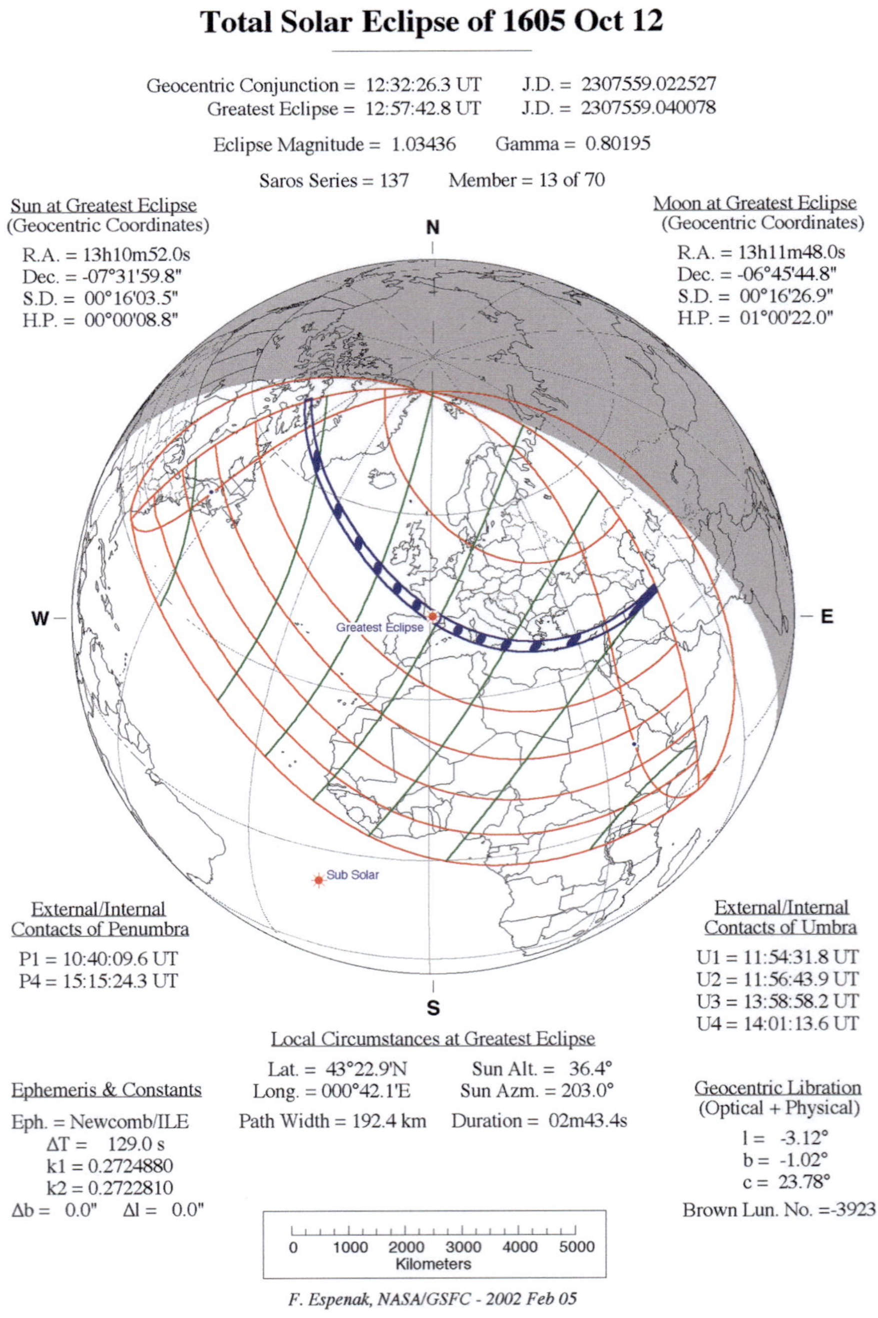

Fig. 3.6 It is still possible to understand the precise trajectory of eclipses that took place centuries ago, thanks to this map prepared by Fred Espenak of NASA's Goddard Space Flight Center. While all England experienced a partial eclipse, the path of totality traveled through parts of Madrid, Spain, on October 12 (2 O.S.), 1605

Fig. 3.7 This view of a thin crescent sun was taken just before totality during the eclipse of 26 February 1979. It also represents what the Sun looked like during the maximum phase of the eclipse of 2 October (O.S.) 1605

Chapter 4
Of Signs and Seasons

Not from the stars do I my judgment pluck, And yet methinks I have astronomy ... (Sonnet 14.1–2)

Part of the ease with which many Elizabethans accepted the tenets of astrology is explained by the exquisite order that the Sun, Moon, and planets show in their courses, an order so delicate it deserves the label of dance. The ballet is represented by the natural and judicial aspects of astrology, a subject familiar to virtually every writer we encounter in this time. It would seem easy to conclude from the opening couplet of Sonnet 14, that Shakespeare was aware of the tenets of natural astrology but did not follow its judicial precepts. But how much of this sonnet represents Shakespeare's real view? At a fundamental level, Sonnet 14 starts with that strong assertion, but it is more of a guideline than a real goal; its opening thought evolves differently with every succeeding line (Sondheim 243–259, Clark 257). The opening two lines seem to confirm Shakespeare's belief that while he understood the aspects of the sky, he did not plan his days, or his plays, according to some perceived meaning of planetary positions in the sky.

Clark cites this as the strongest evidence of Shakespeare's rejection of judicial astrology, that resolution or catastrophic endings of all his tragedies are determined by the characters and the events they originate or in which they are involved, and not by the stars and planets.

However clearly he appears to satisfy his own view in the first two lines, Shakespeare admits later to certain sightings, or omens, in the night sky that would seem to contradict the opening theme, in fact ascribing his future not to aspects of the stars but to those of the young man he addresses. Moreover, in the very next sonnet, Shakespeare claims "that men as plants increase, /Cheered and check'd even by the self-same sky." Both Sidney and Shakespeare sideline astrology "as a poetic image" (Sondheim 249), appreciating all the while the orderly procession that the planets occupy in the sky. It can be argued (weakly) that Shakespeare appears to

D.H. Levy, *The Starlight Night*, Astrophysics and Space Science Library 419,
DOI 10.1007/978-3-319-19878-1_4

reject the judicial aspect of astrology while Sidney accepts it provisionally. This rejection is far from complete, however, and even in sonnet 14 we are left with an indefinite version of Shakespeare's judgment on the causes of major events of his time. In the seventeenth century the distinction between astronomy and astrology, as we now use those terms, was not sharp. We can infer from their inclusion in play after play, by Shakespeare and others, that contemporary audiences enjoyed the astrological content in them.

In early modern thought, *astrologia naturalis* and *astrologia judicialis* were the two accepted ways of studying and interpreting celestial events. The former holds that heavenly bodies have influence on physical manifestations of our lives, like weather and physical matter. The latter insists that these bodies influence human destiny (Sondheim, 243–259). That portion of the old debate survives to this day, though in different form: Eclipses never did affect our futures, but the sky has been shown to have a real connection with human evolution; comets and asteroids colliding with the primordial Earth might well have brought the building blocks of life (Marcus 449). Although Marcus *et al.* base their research upon Halley's Comet during its most recent passage in 1986, the theory's roots lie deep in the astronomical-astrological traditions of the early seventeenth century. Demon stars and eclipses are replaced with comets, collisions, and many observations by ground-based telescopes and spacecraft (Marcus 449).

As a practitioner of science in a time of deeply rooted beliefs in the power of cosmic forces, Johannes Kepler was a pivotal figure in defining the role of astrology at that time; he prepared horoscopes at the same time he was completing observations of the supernova of 1604. His development of the laws of planetary motion was taken seriously throughout England and Europe. Kepler accepted the prevailing maxim of judicial astrology that some events have metaphysical causes. His vision of astrology, however, had little room for the idea of signs, relying instead on his understanding of the simplicity and elegance of the solar system (North 313, 318). Kepler also lacked patience for the spiritualism and demonic magic practiced at the time by such prominent astrologers as John Dee (Casaubon 22, Woolley n.p.), and Lewes Lauaterus (83), who, in his 1572 book *Of Ghosts and Spirits Walking by Night*, revisits the ominous image that "Castor and Pollux have been often seene in battailes sitting on white boxes, valiantly fighting against enemies campe." Beyond the mythological allusion to the battle of Lake Regillus, the twins are also the brightest stars in Gemini, a prominent winter constellation.

There is logic in assigning opinions expressed in the sonnets to Shakespeare himself, just as in Sonnet 26 of *Aristophel and Stella*, Sidney indicates a belief in astrological signs and seasons:

For me I nature every deale doth know,
And know great causes, great effects procure
And know those bodies high, reign on the low.

Sidney's beliefs seem all too apparent here. Shakespeare might "have astronomy" in the natural sense as understood in that time to mean that he accepted the influence of the stars and planets on the four elements.

Early modern audiences probably varied widely in their own astrological judgments; viewers of Shakespeare's theatre shared the enjoyment of readers of *Faerie Queene*, while others took the astrological implications more literally. *King Lear* might have appeared the same year as Bacon's *Advancement of Learning*, but in *Lear*, Shakespeare deposited that anti-judicial astrology rhetoric mostly to the account of Edmund, one of his most malevolent characters. In *Julius Caesar*, it is the conspirators who mock Calpurnia's fear of the repercussion of the stormy night:

> This dream is all amiss interpreted;
> It is a vision fair and fortunate.
> Your statue spouting blood in many pipes,...
> Signifies that from you great Rome shall suck
> Reviving blood, and that great men shall press
> For tinctures, stains, relics, and cognizance.
> This by Calpurnia'd dream is signified. (*JC*.2.2.83–90)

Influence from the stars and planets fills Renaissance literature, not just Shakespeare but most of his contemporaries. It provides one more line of evidence that suggests that astrology was a flourishing subject in the Renaissance; it could be added that astrologers were a common feature of Renaissance courts, but they occasionally appear in government up to recent times; when he was Canada's Prime Minister, MacKenzie King allegedly used both a crystal ball and a Ouija board, and Nancy Reagan frequently called on the services of astrologer Joan Quigley.

A central Jacobean play involving astrology is Webster's macabre tragedy *The Duchess of Malfi (1612–1613)*. Antonio, the Duchess's husband, directs the casting of a horoscope at the birth of their son. In a scene reminiscent of the handkerchief in *Othello*, Antonio loses the horoscope, only to have it found by his enemy Bosola, the man summoned by her brothers to spy on the Duchess:

> Antonio hereabout did drop a paper,
> Some of your help, false friend. O, here it is:
> What's here? A child's nativity calculated! (*Duchess* 2.3.75–78)

The horoscope becomes central to the plot of this play, but its significance is more than that; Webster uses its loss and rediscovery to highlight the play's theme of human destiny vs. happenstance. Lurking in the background is that question, is human destiny predetermined by the stars? In *The Duchess of Malfi*, Webster accommodates the belief of much of his audience that judicial astrology is a real force, but he does not insist that his audiences adopt this view. Indeed, his characters, like those in Shakespeare's *Hamlet, Julius Caesar, King Lear*, and other plays, often present opposing views.

Shakespeare, who was still writing in 1612, offered a similar message. If we go strictly by what appears in his plays, we cannot avoid the inference that Shakespeare took judicial astrology seriously. Edmund's nativity "under *Ursa Major*" would share the spots occupied by Mars and Venus as indicating a "rough and lecherous" disposition, a characteristic that may indeed have been sired by astrological influence, but which has to be borne out in Edmund's behavior particularly at the opening of Act II. It may well be that according to Ptolemy's thinking, Edmund's role in

Lear's subplot is predetermined, whether by the positions of Mars and Venus in the night sky or by that of Ursa Major (Rusche 163), but more likely by the twists of fate of Shakespeare's pen. Edmund begins by quoting his own horoscope; after being conceived under a portion of one constellation (Draco's tail) he was born under Ursa Major, the Greater Bear. Neither constellation is even close to the zodiac where characters figure in all births.

As humorous and absurd as it appears now, the complexity of judicial astrology means that the mere existence of houses for Mars and Venus condemns Edmund to a sorry existence. However, had Draco's tail not been involved, Edmund could have emerged a reasonable man. These evil astrological omens shaped his "rough and lecherous" demeanor, but Edmund disregards the entire horoscope with his very next word "Fut!" Tossing out the entire astrological approach to his behavior, he insists instead that "I should have been that I am had the maidenliest star in the firmament twinkled on my bastardizing." (*KL* 2.1.127–129) Edmund strengthens that view in his argument with Gloucester about the alleged predictive powers of eclipses (1.2.110–117).

Turning Observation into Theory

While Galileo was not the first to use a telescope, he was almost certainly the first to acquire a large dataset of observations, and then publish them widely. However, if Shakespeare knew of the astronomical works and theories of Thomas Digges, then Peter Usher's contention that Hamlet contains allegorical hints about Copernican astronomy would be worth investigating. Usher's idea that the entire play is an allegory pitting the Copernican theory against that of Ptolemy is unconvincing, but Usher does make good points regarding some details of Hamlet's troubles. For example, Laertes uses the term "blastments" in *Hamlet* 1.3.42. Although this word refers to wounds on plants, Usher extends its meaning to the large craters on the Moon that were blasted out billions of years ago by impacts from passing asteroids or comets. If this is Shakespeare's intended meaning, which I doubt, then the playwright has predated Eugene Shoemaker's impact theory by almost four centuries (Shoemaker, 70–89). In our time, not only have we walked across small craters of the Moon, but we have witnessed also the process by which craters on terrestrial planets can be formed (the collision of a comet with a planet in 1994), and in 2005 we experienced the actual making of a crater on a celestial body when the Deep Impact spacecraft collided purposefully with Comet Tempel 1 (Levy 218).

As we have seen from Edmund's answer to his father's fear of the "late eclipses", Shakespeare was aware of the official sanction against judicial astrology. There are other crucial references to the alleged power of the cosmos over human destiny in later plays, such as Cassius's plea in *Julius Caesar*, "The fault, dear Brutus, is not in our stars/ But in ourselves …" (1.2.139–141), and Macbeth, whose anticipation of his ill-gotten royalty led him to command the stars to "hide your fires," rather than being commanded by them (1.4.50–51). The distinction between astronomy and astrology would grow after the early use of the telescope in 1610, but throughout most of Shakespeare's career it was not very pronounced.

> In *Troilus and Cressida* (circa 1602) Ulysses describes (1.3.85–94) how
> The heavens themselves, the planets, and this centre,
> Observe degree, priority, and place … (*TC* 1.3.85–89).

"This centre" seems deliberately vague; it could refer to Earth, as in the Ptolemaic cosmic view, or to the Copernican Sun. Shakespeare typically avoided discussing cosmic theory in his plays, possibly out of consideration that his audiences had come to see a play, not a tome. Moreover, although his leading villains often ascribe their courses of action to astral influence, not a single one of Shakespeare's plays achieves its resolution directly from the stars and planets, even if the characters believe they are so controlled, with the possible exception of the misbehaving heavens the night before Caesar's assassination in *Julius Caesar*. I make this statement despite the interpretation common among scholars that in most Shakespearean tragedies, the protagonists fail because of a combination of their own failings and cosmic forces. However, in virtually every Shakespearean Tragedy the protagonists' fates result partially from circumstance, which they seem often to attribute to the fate of the stars and planets; i.e. judicial astrology. These characters claim influence from among three types of astrological influence; expressing a mood like the comets at the opening of *1 Henry VI*; as a symbol of elation or misfortune, as in *Titus Andronicus* whose protagonist, like Don Quixote chasing windmills, shoots arrows at the planets after his betrayal by his friends and colleagues (4.3.62–75); and as direct evidence for or against astrological *power*. In the vast majority of examples, like the "huge eclipse" in *Othello* (5.2.100–102), the star-crossed love between Romeo and Juliet, and the "five moons" traditional prophecy of royal death in *King John*, the "evidence" turns in favor of that power. In Shakespeare's comedies, particularly *All's Well that Ends Well*, astrology has the additional role of adding humor, as in the Clown's remark about "a good woman born before every blazing star, or at an earthquake" (*AW*.1.3.81–83).

Shakespeare filled his plays with the kinds of characters his audiences wanted to see; witches and faeries that made astral prognostications come true, evil characters who denounced astrology and who also practiced the cold and demanding rules of modern science and logic (Clark 257). The conflict over double meanings of eclipses in *King Lear*, whether they be portents or scientific events alone, heralds the larger issue of how Nature works in this play. The eclipses mirror a larger argument about the doctrine of natural law that claims Nature as being moral versus the Machiavellian-Hobbesian credo that Nature is amoral (McAllindon 163).

It is not really possible to come to a definitive conclusion about Shakespeare's own belief, notwithstanding his own statement in Sonnet 14's first line. Like most of his contemporaries, Shakespeare presumably accepted the order presented by natural astrology (Clark 258–259). He enriched his stories with axioms from judicial astrology even though, based on the first line of Sonnet 14, he probably did not follow them.

Replete with end-of-world predictions, Richard Harvey's *Astrological Discourse* threatened a Noachian flood-like "great abundance of waters" and other misfortunes reminiscent of Gloucester's later prediction in *Lear*. Despite Harvey's conformity with traditional astrological views of the time, his efforts were met with ridicule and scorn from the public, and more privately from his brother Gabriel, whose unconvinced response led to a more tentative pamphlet 5 years later entitled *A Discoursiue probleme concerning prophesies*, which lists many examples of "successful" forecasts. Long on perceived past success but short on prediction, such kinds of writing persist to the present day.

It is interesting to see how some of the best early modern literature, by authors as diverse as Shakespeare, Spenser, and Davies, combines its treatment of contemporary astrology with bold looks into the major developments in the night sky. If Usher's hypothesis is correct, then Shakespeare was far more aware of the night sky than I give him credit for. I see *Hamlet* primarily as a play intended for a contemporary audience, but its story can be read as allegory. In Usher's words, Hamlet personifies Copernicus, Claudius harks back to the Ptolemaic Earth-centered system, and Rosencrantz and Guildenstern reflect Tycho's compromise. I doubt that any telescope built prior to the middle seventeenth century would have been good enough to allow Shakespeare to view details as small as Jupiter's Great Red Spot, and it is in such details that Usher's argument fails. Nevertheless, the evidence Usher offers does outline a rich body of knowledge in astrology, where characters and plots reflect an increasing comprehension of the night sky of their time.

Astrology and the Digges Family

That Shakespeare's reading in astrology was sophisticated comes from considerable circumstantial evidence. His lodgings in London were not far from those of Thomas Digges, one of the most learned scientists of his day, and from the home of Elizabeth's "science courtier" John Dee. Of previous investigators who have tried to portray the works of the Digges family, Francis R. Johnson is probably the best known. However, a few clarifications to his classic 1938 paper on Copernicus are in order (Johnson 390–410). Johnson's first sentence assigns the beginning of modern astronomy to Copernicus. Actually this is true for modern astronomical theory, but the equivalent in modern observational astronomy dates back to Tycho Brahe, whose observations of the Supernova of 1572 were as precise as the instruments of his time would allow. In England, Thomas Digges was right behind Tycho Brahe in reporting the supernova of 1572 (Johnson 391).

Notwithstanding the probability that Leonard Digges (Thomas's father) constructed a sixteenth century telescope (see Chap. 5), prior to 1610 night sky observations were generally accomplished and recorded without optics. Galileo's telescopes would spark a basic alteration not only in what objects could be studied, but also in how they could be studied. In 1605, there were essentially no telescopes, although it would not have been impossible to imagine Thomas Digges using some modification of his father's periscopic lens, actually the major portion of a small telescope, to view the October solar eclipse over London. After 1610, contemporary ideas in judicial astrology would add four new Jovian moons to keep up, along with the changing positions of the signs over time due to precession of the Earth, the result of the planet's 22,500 year wobble. The telescope offered a renewal and an expansion of humanity's understanding of the cosmic neighborhood, but one that astrologers have generally not taken advantage of in the last 400 years. The reason for this is that astrologers have not generally paid attention to precession, the result of Earth's wobble, like a slow-moving top, once every 22,000 years. Precession was understood in early modern times. Spenser ambiguously describes it in *Faerie Queene* V (*proem*.5.6–9):

> For that same golden fleecy Ram, which bore
> *Phrixus* and *Helle* from their stepdames feares,
> Hath now forgot, where he was plast of yore,
> And shouldred hath the Bull, which fayre Europa bore.

Spenser's remark on mythological history implies his awareness that, some 1500 years after Ptolemy, the constellations had shifted approximately one sign eastward, so that the "fleecy Ram … hath now forgot" his former place. Someone traditionally born under the sign of Ram, for example, is technically now under Taurus the Bull. The major theme of *Octavia*, a work Renaissance writers thought had been written by Seneca but now doubted as such, was a catastrophic war among the stars, reaching its disturbing climax as all the stars fall from heaven, including a final deathlike setting of the Sun.

The English literary Renaissance is hardly trivial compared to Seneca's golden age, but Spenser takes the war among the stars of heaven and recasts it in terms of precession (Bull, 419); it can be claimed that by the 1590s, Spenser had replaced classical mythology with positional astronomy (Bull, 417–419). Because of this interpretation, Spenser's lines are enhanced by his modern view of a developing interplay between science and literature. In his telling of the legend of Artegall, Spenser portrayed "That all the world with goodnesse did abound" in the Golden Age during Saturn's rule (*FQ* 5 *Proem* 9).

In about 1576, well within Spenser's time, Thomas Digges proposed the extraordinary idea that space is infinite. This claim precedes by 7 years the better known claim by Giordano Bruno, who proposed independently the idea of the infinity of space in 1583, and who paid for it with his life at the stake in 1600. This makes Spenser's claim that if all the stars were all to fall from heaven at the time of judgment, then our own star, the Sun, would be among the fallen along with all its worlds. This thought adds a miniscule amount of scientific credence to the idea of the falling of *all* stars, including the Sun. Beyond the orbits of the planets, the stars are increasingly distant from the solar system, spreading out to infinite distances. The idea that not the stars, but the farthest superclusters of galaxies with their stars, might hover at the edge of infinity might be Digges's most important original contribution to knowledge, as well an influence upon the literature of the time.

Sir Henry Wotton also used poetic allusions to the night sky of that time. Although his poetic output was small, it did offer such playfully probing questions as this poetic note flattering King James' daughter:

> You common people of the skies,
> What are you when the sun shall rise?
> (Wotton, "You Meaner Beauties of the Night" 1.4–5)

Marlowe's *Tamburlaine* might be set up by the astrological design of a poorly housed planet like Jupiter, but it ends as a "tragedy of inordinate passions" built not from stars and planets but from the author's development of his plot and characters. The placement of Jupiter might help the plot along, not the reverse. Tomkis's *Comedy of Albumazar* (1614) offers astrology as a virtual co-conspirator with thieves in its narrative, beginning with the smallest planet:

> Your Patron Mercury is his mysterious character,
> Holds all the makes of the other wanderers,
> And with his subtill influence works in all
> Filling their stories full of Robberies … (1.1.11–14)

Simple and elegant descriptions of dusk, dawn, sunrise, the Moon and Sun, and planets offer a window of opportunity to interpretation. Francis Bacon, who in 1605 was developing his modern scientific method, was producing a new kind of reasoning based upon observation rather than on philosophical intuition. Wotton notes the daily fading of the stars as sunrise approaches in this poetic letter to Elizabeth of Bohemia. Edmund offers a modern notion of eclipses in *King Lear* before his father, who detailed his more traditional astrological views in the eclipse scene (KL 1.2), has his eyes ripped from him. Romeo admires a morning sky where Juliet shines above all as the Sun (*RJ*.2.2.1–3). Daybreak is a call to action to assassinate in *Julius Caesar* (2.3.75–115) and to discover that an assassination has occurred (*Macbeth*.2.3.61–78).

In comedies, dawn and sunrise typify the most optimistic parts of a day, as they did in Fletcher's *The Woman Hater* offering another example of a hitherto unappreciated, thematically significant allusion to a celestial phenomenon visible every morning. Entered into the *Stationer's Register* in 1607 not long after the supernova of 1604 and the eclipses of 1605, this play begins with an elaborate metaphorical description of the constellation of Virgo rising in a predawn sky. At the lateness of the hour on a December morning, the Duke wonders

> is it so much, and yet the morn not up?
> See yonder where the 'shame-faced maiden comes
> Into our sight, how gently doth shee slide,
> Hiding her chaste cheeks like a modest Bride,
> With a red vaile of blushes. (1.1.3–8)

The predawn rising of Virgo is a form of celestial time-keeping that can track the passage of time in that play, and when connected to the earlier note of the actual time of day as past four in the morning, becomes also a way of synchronizing the celestial clock with artificial timepieces. Since Virgo rises at dawn only at one particular time in the year, i.e. December, that event is an indicator also of the time of year of a particular incident. In *Woman Hater*, that wording is not some vague portrayal of dawn, but a specific picture of the appearance of Virgo rising in the southeastern sky as dawn begins. I suspect that Fletcher actually observed the predawn scene, or at least interpreted it from someone who had seen it. Virgo was low in the southeastern sky "past four" a.m. in mid-December 1605 and at the same time in 1606. Virgo is not a conspicuous pattern of stars. Besides its brightest star Spica, the figure is "shame-faced" and unimpressive, even more so with interference from dawn. Why would an observer's attention be drawn to such a sight? On the morning of 15 December 1605 as Virgo rose, the bright red planet Mars was about a Moon-diameter from Spica, forming a pair of bright "eyes" that easily drew attention to the constellation. One can be fairly certain that both the Virgo reference, and the time mentioned as 4 o'clock means that if Fletcher actually saw Virgo in the predawn sky, he had to observe it near dawn on a mid-December morning. The Moon's orbit around the Earth, on the other hand, offers a timepiece that measures either the passage of months or the particular time of a month. Spenser uses this clock of lunar phases at least three times in *The Faerie Queene*:

> Now haue three Moones with borrow'd brothers light,
> Thrice shined faire, and thrice seem'd dim and wan (*FQ* 3.3.16)

Both passages measure the passage of 3 months;

> But till the horned moone three courses did expire (4.6.43)

also measures 3 months that begin at a crescent phased Moon; and finally

> As the faire Moone in her most full aspect (5.5.3)

defines a full Moon. These lines contrast nicely with those of Sir Walter Raleigh, whose tribute to "Diana's fair and harmless light" sees the Moon not as a means to an end but as an object of beauty by itself, a queen of the sphere of night.

During the 1580s and 1590s Tycho wrote and published his most noteworthy lines from his observatory about the role of judicial astrology in the interpretation of astronomical phenomena relating to the star of 1572: "Moreover, forasmuch as this Starre was placed in the eighth Spheare, above the Orbes of the Planets, it seemeth that the predictions issuing from it, do not only concerne one peculiar tract of Land, but all the Nations of the world; and therefore it will bee the longer before the effects will be declared by succeeding events." He suggested that a "Sybils Pophesie" dealing with the end of the world would result from the appearance of a star in the north (the supernova in Cassiopeia) that causes nations worldwide to "lay by their weapons and imbrace peace" (Brahe, 22). In contrast to this seeming good news, the star would be followed by a bellicose comet "with martiall sparkling beames" ushering in a more difficult era.

The Great Conjunction of 1583

The complex terminology of judicial astrology, with its conjunctions and trigons, found its way into Shakespeare without any obvious contemporary reference except for a view of the two planets, Jupiter and Saturn, slowly closing in on each other early in 1583. The practice of astrology dates as far back as biblical times (for signs and for seasons, *Genesis* 1: 14). During the early modern period, a basic level of astrological learning was accepted as general knowledge, and a lack of awareness of every nuance would not have prevented a contemporary audience from enjoying the more obvious allusions to the stars (Aston, 160). Iago complains about Cassio and Montano launching "in opposition bloody/as if some planet had unwitted man," simply acknowledges that the stars could offer a guideline for human behavior, without bringing forth any concrete evidence that they actually do so (Sondheim 246).

In astrology the signs divide into trigons, three for each of the traditional "elements" of earth, air, fire, and water. An ordinary conjunction invokes a narrowing of the apparent distance in the sky between a planet and the sun, but a "great conjunction" is historically defined as one in which Jupiter closes on Saturn in the sky. Such an episode occurs only once in approximately every 20 years. One such event took place in 1583. Astrologers found it important to know in which constellation such a conjunction took place. If the time of conjunction, or when the two planets are closest to each other, occurs when they are easily visible in the night sky, then determining the constellation through which Jupiter overtakes Saturn is simple. But if the actual conjunction occurs when the planets are so close to the Sun in the sky that no one actually sees them, it is far more challenging. Most Elizabethan astrologers considered these conjunctions as critical events, the most notorious of which took place in 1583 near the boundary between Cancer and Leo and which led to Richard Harvey's *Astrological Discourse*. Astrologers called Leo one of the signs of the "fiery trigon" whose other two signs are Aries and Sagittarius. The conjunction of 1583, they noted, would be the last for centuries in any of the "watery trigon" constellations. Though the possible effects of the great conjunction were well known to astrologers, they were not impressed upon typical English readers until early 1583, when Harvey's *Discourse* announced the conjunction and foretold of many consequences, such as huge upheavals and even a "Last Judgement" in 1588.

On the date Jupiter and Saturn were closest; both planets were so close to the position of the Sun in the sky that neither could be directly observed. As a result, a dispute arose as to whether the conjunction really took place in the watery trigon sign of Pisces, or the first of a series to occur in a fiery trigon sign, like Aries. Bringing his great observing skills to bear, Tycho made meticulous observations of Jupiter and Saturn, and confirmed that the conjunction had indeed taken place in Pisces, though close to its border with Aries. The conjunction of the two giant planets that followed in 1603 was well within the fiery trigon.

So much was written about trigons that the term became part of the general vocabulary of the time. Astronomer Owen Gingerich notes, for example, that in the tavern scene of *2 Henry IV*, Prince Henry and his comical friend Edward Poins compare Falstaff's tryst with Doll to that of two planets. "Saturn and Venus this year in conjunction?" asks the Prince. "What says th' almanac to that?" We can actually answer that question using contemporary sources like Harvey's *An Astrological Discourse*. Poins adds, "And look whether the fiery Trigon, his man, be not lisping to his master's old tables, his note-book, his counsel-keeper" (*2 Henry IV* 2.4.253–257).

The fiery Trigon is Shakespeare's clever allusion to a red-faced Bardolph, who makes advances to the tavern's hostess Mistress Quickly. Shakespeare, who wrote this play around 1598, might have had in mind the great conjunction from 5 years earlier, and the one to come in 1603 (Gingerich 394–395).

J. W. Draper supports the idea that many people were committed to these ideas, adding that Queen Elizabeth herself regularly employed John Dee to compute for her the lucky hours and days for conducting her affairs (Draper 20). Defying royal custom, Elizabeth observed the Great Comet of 1577, and Dee was considered a Copernican astrologer whose advice was sought more for its informative value than for its predictive power. In any event, at the time there was considerable religious opposition to judicial astrology, since its practice was considered by some officials to be a violation of the first of the Ten Commandments (*Exodus* 20.2). According to Warren Smith, the idea that Elizabeth's second parliament passed a series of anti-judicial astrology bills, each one harsher than its predecessor, indicates that part of the British population did practice astrology and that another portion objected to it. In 1550, Smith points out, "Bishop Hooper says belief in astrology is against the first commandment" (Smith 160). Elizabeth's Act of 1580, in fact, compared judicial astrology to witchcraft and threatened the death penalty for the practice of either. James I also disdained the practice, distinguishing between *astronomia*, the law of the stars, and *astrologia*, the preaching of the stars. Of the latter, James wrote: "It is this part which I called before the devils schole" (Smith 159–176).

However, none of this prevented Shakespeare and his colleagues from dramatizing the relationships between character and situation. We can never know what Shakespeare's stand on astrology was, if he really had one, but he surely was aware that his audiences shared a belief in its power, and that acknowledgement manifests itself differently in each tragedy. I doubt that anyone in an Elizabethan theatre audience would have been scandalized by the announcement in the Prologue that Romeo and Juliet were "star-cross'd lovers; (*RJ*.6)"; Cassius's famous remark that "The fault, dear Brutus, lies not in our stars, but in ourselves, that we are underlings" (*JC*.1.2.140–141) contrasts nicely with Calpurnia's equally celebrated warning

> When beggars die, there are no comets seen
> The heavens themselves blaze forth the death of princes. (*JC* 2.1.29–30)

Seen against this historical background, Shakespeare's frequent allusions to astrology would be expected to raise questions in his audiences. In *The Two Gentlemen of Verona* (circa 1594), Julia asserts the influence of the stars on her own destiny when she speaks of her love for Proteus (2.5.73–75). *Romeo and Juliet*, from about a year later, has so many astrological allusions that Shakespeare might have intended to suggest that astral influence was a cause of this tragedy of two "starcross'd lovers" (*RJ*.Prologue.6), or to deepen and problematize the tragedy by infusing it with so many vexing astrological markers. Mercutio has the temper of Mercury, for instance, and the Friar, who represents the religious thought of "for signs and for Seasons", (*Genesis* 1.14.) blames Capulet's misfortunes on the stars (4.5.9495). The Captain in *Richard II* (1595) asserts the judicial power of the cosmos (2.4.711), and *Hamlet's* Horatio (circa 1600) recalls a scene from the earlier *Julius Caesar*, then adds further lore to it (*Hamlet* 1.1.114–120). Both passages allude to meteors, parenthetically "meteors fright the fixed stars of heaven" (*Richard II* 2.4.9) and "stars with trains of fire" (*Hamlet* 1.1.117). These allusions were too early to relate to the great Leonid meteor storm of November 1602 (which was not reported) over England. Dee's diary notes describe a strange meteor-like object appeared on the evening of 26 August, 1581:

> Aug. 26th, abowt 8½ (at night) a strange meteore in forme of a white clowde crossing galaxiam, whan it lay north and sowth over our zenith; this clowd was at length from the S.E. to the S.W. sharp at both endes, and in the west ende it was forked for a while; it was abowt sixty degrees high, it lasteth an howr, all the skye clere abowt, and fayr starshyne.
>
> (Dee, http://www.gutenberg.org/files/19553/19553-h/19553-h.htm)

If it were a meteor, then it had a train that lasted for almost an hour; otherwise it might doubtfully have been a long-lived ray from a display of *aurora borealis*.

In 1594 Sir John Davies' *Orchestra, or a poeme of dauncing* combined actual observing of the night sky with its philosophy of interpretation, as a challenge to the magnificent cosmic dance he envisaged. In order to follow this dance the reader needs to understand the details of the individual pattern of the Moon as it orbits the Earth. In this stanza, conflicting philosophies are in agreement:

> Who doth not see the measures of the Moon
> Which thirteen times she danceth every year?
> And ends her pavan thirteen times as soon
> As doth her brother, of whose golden hair
> She borroweth part and proudly doth it wear.
> Then doth she coyly turn her face aside,
> That half her cheek is scarce sometimes descried. (*Orchestra* 25)

Davies' poetic description of the Moon's phase needs no scientific change after four centuries. He correctly notes that the Moon always points the same face toward Earth; she faces him at full phase, then turns her face aside as the phase wanes. Would that the rest of the dance be so easily described! Davies cannot sing about the planets so easily:

> Only the earth doth stand forever still:
> Her rocks remove not, nor her mountains meet;
> (Although some wits enrich'd with learning's skill
> Say heaven stands firm and that the Earth doth fleet
> And swiftly turneth underneath their feet)
> Yet, though the earth is ever steadfast seen,
> On her broad breast hath dancing ever been. (*Orchestra* 51)

Davies gives the new ideas three lines in parentheses, and then concludes that the Ptolemaic Earth-centered system fits the order better. The important astronomical theme of *Orchestra* is that celestial bodies are dancing in a pattern dictated by gravity, of the Sun, and of its kingdom of worlds. Dance is echoed throughout the major writing of this period, from the dance fantasies in *A Midsummer Night's Dream* to the cosmic dance in *Orchestra*. Far more than a simple declaration that all's well in the world, the dance theme suggests, in a time of scientific uncertainty, a sense of great order in the sky that one can see each night. It is an order that, as

James Miller wrote, “could pacify the tempestuous world of matter by inducing a chorus of airy spirits to dance out the mysteries of world-harmony in the shadows of the cave, but strove instead to heal humanity’s distempered soul with celestial music” (Miller 465) (Fig. 4.1).

Fig. 4.1 Sunset from Jarnac Observatory. This late fall sunset depicts one of the most fascinating attributes of the sky—a beautiful sunset in a clear sky that heralds a perfect night. Photograph by the author

Chapter 5
The Telescope in Early Modern English Literature

Some 40 years ago the renowned cosmologist and observatory director Harlow Shapley wrote of Job's interrogation by G-d from Chap. 38 in the biblical *Book of Job*. "This is no elementary quiz," the great astronomer wrote. "I would call it a swift-moving doctoral oral" (Shapley 143). Shapley gives this ancient parable a modern interpretation of a tortured man struggling to understand his relation to the Universe. "Were you there," asks G-d, "when I created the stars of the Pleiades or Orion?" The birth of a star is one of the most beautiful and violent processes that our galaxy offers: A long period of dark, impenetrable cloudiness (the specific prenatal cloud is called a Bok globule) is followed by an ignition flash as the nascent Sun begins nuclear fusion. At the end of the process, the new star's surrounding nebulosity quickly burns away.

The telescope introduced a sense of democracy to the night sky. The early telescopes were not difficult to fabricate, and anyone with access to lenses and a tube could make one. In those days, the sky visible to the unaided eye offered much more than it does through light-polluted air we have at present. However, the telescope also served to create an "elite" kind of "observer class" of astronomers who could see objects and events beyond what the unaided eye can see. As we comprehend it today, that Universe is vastly different from the idea of stars extending into infinity that Thomas Digges first proposed in 1576. Giordano Buno may have been executed for it, but the Englishman Digges thought of the notion years earlier in a land whose religious heritage did not systematically and aggressively oppose new theories and discoveries. In his *Perfit Description*, Digges's expanded translation of Copernicus's *De Revolutionibus* that exists as part of the English author's *Prognostication Everlastinge*, it was the stars in our visible sky, in our own galaxy, that stretched on endlessly. It is understood currently that the Universe, not the Milky Way galaxy, is infinite, and that its great superclusters of galaxies, not the stars within them, stretch in perpetuity. But in a major way, Digges was on the right track when he wrote as explanation for the space beyond the orbit of Saturn: "This orbe of stares *fixed infinitely up* extendeth hit self" begins his lengthy and detailed caption regarding "the last and highest of all; the immoveable sphere of the starres …" (emphasis added, T. Digges (1576)).

D.H. Levy, *The Starlight Night*, Astrophysics and Space Science Library 419,
DOI 10.1007/978-3-319-19878-1_5

I do not see any conflict between Digges' idea of the sphere of stars extending to infinity and John Dee's theoretical calculation of the actual distance to that sphere. In his *Preface to Euclid* (published some years after his death in 1605), Dee writes that "the distance of the Starry Skie is, frō us, in Semidiameters of the Earth 20018 (and an illegible fraction)". And to ensure no one missed it, he follows this numerical statement with a repetition in words: "Twenty thousand fourscore, one and almost a half." (Dee, n.p.) The beginning of the "starrie sphere" is only that far, but that beginning does not preclude the stars within the sphere extending on to infinity. I assume that this inner boundary of the most remote sphere was indeed considered fixed at the distance calculated by Dee.

In addition to being at the center of our local system of planets, our Sun sits in the center of our region of the Universe because there is no "better or conueient place than this, from whence vniformely it might distribute light to all, for not vnfitly it is of some called the lampe or light of the world (Digges 359)." If, as Digges goes on, the Universe is infinite, then there can be no center, or rather every point within it can be seen as the center. Thus, the orb of the stars contains "a central, immotile, finite symbol of temporal power." The stars are natural homes of G-d, as there they are the most "conuenient place … from which vniformly it (they) might distribute light to all, for not vnfiutly it is of some called the lampe or light of the world." In this sense, Digges's theory of infinite space has no contradiction with Troilus's words:

> The heavens themselves, the planets, and this centre,
> Observe degree, priority, and place … (*TC*. 1.3.85–94)

This statement is commonly interpreted with the centre being the Ptolmaic Earth, but it works equally as well if "this centre" is the Copernican/Diggesian Sun as a central star, even justifying the possibility with Troilus's proviso that, in the years since Copernicus, "The bonds of heaven are slipped, dissolved, and loos'd" as they certainly would have appeared to be in the ake of Galileo's discoveries in 1610 (*TC* 5.2.154, Usher 2).

In its earliest form, Copernicus's theory did not predict planetary positions much more accurately than did the established system of Ptolemy. Kepler's modifications to that theory did however; in 1605 Tycho's former student introduced three laws of planetary motion that specified that the planets move in elliptical orbits, not circles, with the Sun at one focus of the ellipse. The advent of the telescope in late 1609 and 1610 confirmed these ideas observationally, and ushered in an age of enormous scientific advance.

After thousands of years watching the planets move obediently in their courses, the early telescope opened a universe incredibly bigger and more interesting than the one with which the world was familiar. That opening began in January 1610 when Jupiter reached opposition from the Sun and was brightly visible in the evening sky. Within less than a week, Galileo reported four small "planets" accompanying Jupiter. Not specks of light but large worlds, these moons of Jupiter presaged more scientific advance in an instant than at any other time in history.

In Italy, British ambassador Sir Henry Wotton hastily arranged to get an early copy of Galileo's *Siderius Nuncius* to King James's chief advisor and close friend,

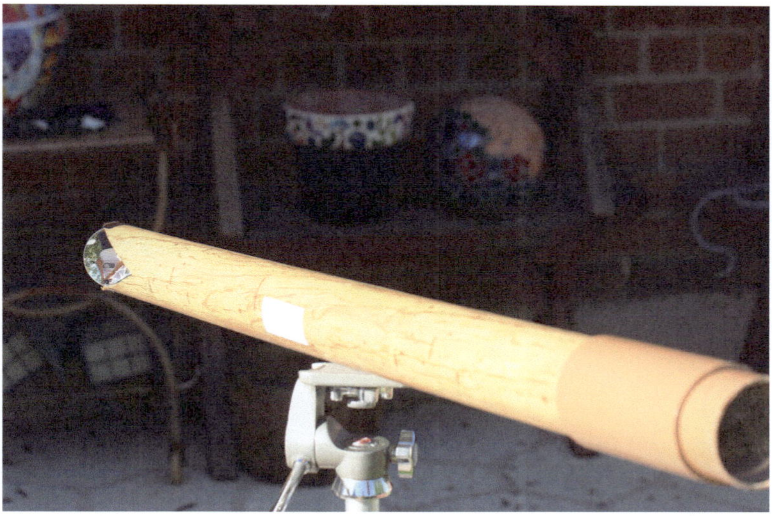

Fig. 5.1 Ewen Whitaker built this perspective lens instrument, which he credits to Thomas Hariot, William Bourne, and Leonard and Thomas Digges. Photograph by the author

Sir Robert Cecil. The copy was accompanied by this enthusiastic letter: "I send herewith unto his Majesty the strangest piece of news (as I may justly call it) that he hath ever yet received from any part of the world … he (Galileo) hath first overthrown all former astronomers" (678). Obviously intrigued and fascinated, Wotton wanted to spread the news of Galileo's discoveries as quickly as possible. The difference was that Copernicus's mid sixteenth-century work was theoretical, but Galileo's discoveries were based entirely upon his personal experience of direct observation of the night sky.

"With this instrument we can descry those small stars wheeling as in a dance round the planet Jupiter," wrote Francis Bacon in *Novum Organum*, "whence it may be conjectured that there are several centers of motion among the stars" (Bacon 193).

The New Atlantis, Perspective Lenses, and the Night Sky

The New Atlantis was published in 1624 in Latin, and in English 3 years later. In it, Bacon writes:

> We have also perspective houses, where we make demonstrations of all lights and radiations and of all colors; and out of things uncolored and transparent we can represent unto you all several colors, not in rainbows, as it is in gems and prisms, but of themselves single. We represent also all multiplications of light, which we carry to great distance, and make so sharp as to discern small points and lines. Also all colorations of light: all delusions and deceits of the sight, in figures, magnitudes, motions, colors; all demonstrations of shadows. We find also divers means, yet unknown to you, of producing of light, originally from divers bodies. We procure means of seeing objects afar off, as in the heaven and remote places;

Fig. 5.2 This is a surprisingly wide field view of a distant tree seen through Dr. Whitaker's perspective lens. Apparently it works as advertised! Photograph by the author

and represent things near as afar off, and things afar off as near; making feigned distances. (Bacon, *New Atlantis*, 110)

With this passage Bacon considers perspective glasses on a different level. Adding to the body of English literature that describes what these lenses can accomplish, he insists that "we procure means of seeing objects afar off," (Bacon 110) *specifically* writes "in the heaven and remote places." These glasses can show "things afar off as near," just as a telescope would later do. This appears to be the first positive indication that perspective lenses were used, or at least could be used, to observe the night sky; I find no evidence that earlier researchers noticed this intriguing sentence.

John Donne was probably the first English author to weave the discoveries of 1610 into his *Anatomie of the World*, where

Tis all in pieces, all cohaerence gone;
All just supply, and all Relation:
Prince, Subject, Father, Sonne, are things forgot,
For every man alone thinks he hath got
To be a Phoenix, and that then can bee
None of that kinde, of which he is, but hee. (*Anatomie*, 205–218)

As long as the Sun-centered solar system was only theoretical, there was little need to consider it. But Galileo's discoveries were real and could be viewed and confirmed by anyone with a telescope. Donne uses the death of a single person to

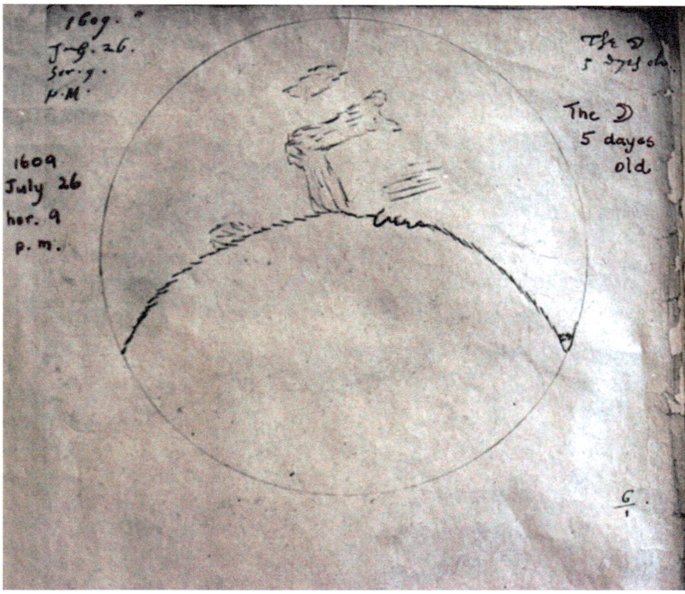

Fig. 5.3 Thomas Hariot's drawing of a 5-day old Moon, on 16 July 1609, preceded Galileo's drawing by a few months. The author took this photograph from Dr. Whitaker's own images, which he in turn obtained from Hariot's manuscript in 1970. In this single image, as in the next, we almost physically travel back in time to the very earliest observations made with a telescope

present a dying old world. The subjects of royalty and family are forgotten because they are familiar subjects, while the new astronomy, correct as it might have seemed before 1610, seemed now "incoherent" and uncomfortable to fathom. Every man must try to figure it out independently (to be a Phoenix), Donne insists. He offers a personal interpretation of the new philosophy, beckoning it to the theme of utter bereavement that begins his elegy.

The evidence that telescopes were around long before 1609 does not necessarily go as far back as Democritus, whose opinion that the Milky Way is composed of large numbers of stars implies that this Greek scholar actually viewed the sky through a telescope. However, I have seen the Milky Way appear mottled, as if comprised of many stars, on many dark nights lacking any optical aid. Without telescopes, I suspect that some expert observers of old had exceptionally keen eyesight, not to mention an unpolluted night sky. "Shakespeare lived in a world of Time, Milton in a universe of Space," wrote David Masson in 1966. "It happened as suddenly as that." (Devereux, 91) Moreover, Shakespeare alludes to the evolving new technology in several plays. He surely was capable of noticing the mottled appearance of the Milky Way both from his native Stratford and from London.

The telescope's arrival around 1609 was not without warning; it was preceded by a technology called perspective or proportional glass that was formed in such a way that from a particular vantage point was capable of forming a magnified image of a distant object. The technology evolved over some 40 years and was a logical precursor to the telescope. The Digges family reckons prominently among those who experimented with early versions of perspective glasses that could function as telescopes. A modern telescope mirror used without an eyepiece actually functions as a perspective glass. It makes sense to credit Leonard Digges with autonomously inventing the telescope, as his surveying career required him to measure distant objects.

It is important to note that the telescope was used in England before 1610 as the perspective glass, and after 1610 as Galileo's "optic tube." Shakespeare's lodgings in London were not far from those of Digges and Dee, and he might have known them personally, as Owen Gingerich suggests (Gingerich 395). That the Bard knew of their work, however, is clearly shown by his allusions to it in three of his plays; *Richard II*, *Henry V*, and *Hamlet*, where Shakespeare alluded to the implausibility of the idea of infinite space in Hamlet's argument to Rosencrantz and Guildenstern: "O God, I could be bounded in a nutshell and count myself a king of infinite space, were it not that I have bad dreams" (*Hamlet* 2.2.253–256).

The expression "proportional glass" implying glass capable of enlarging or shrinking views of objects, appears in Leonard Digges' Pantometria, published posthumously by his son Thomas in 1571, 39 years before Galileo's *annus mirabilis* of 1610. After celebrating the "resplendent heavenly Globes of Sunne, Moone, Planets and Starres fixed," Digges reveals that "my father by his continual painful practices, assisted with Demonstrations Mathematicall, was able, and sundrie times hath by proportionall Glasses duely situate in convenient angles, not onely discovered things farre off" … Moreover, the elder Digges "by the Sunne beams fired powder, and discharge Ordinance half a myle…." In 1579 Thomas Digges completed his father's *Stratioticos*, and wrote further that his father "was able by Perspectiue Glasses (glasses apparently able to show objects differently depending on how they are used) duely situate upon convenient Angles, in such font to discover euery particularitie of the Country round about, wheresoeuer the Sunne beames might pearce; as whence Archimedes (Bakon of Oxford onely excepted) I haue not read of any in Action euer able by means natural to performe the like" (Digges 359). Digges's friend, mentor, and surrogate father John Dee, doubtlessly also became interested in perspective glasses: "… if you, being (alone) near a certaine glasse, and proffer, with dagger or sword, to foyne at the glasse, you will suddenly be moued to giue back (in maner) by reson of an image, appearing in the ayre, between you & the glasse…."

Dee unabashedly credits himself with the invention of "Proportionall and Paradoxall Xompasse (sic) (of me inuented, for opur two Moscouny Master Pilotes)"…. (Dee n.p.) Though he is probably correct in that these glasses were first used in England, Dee was but one member of a team that included the Digges father and son, Leonard and Thomas.

The fact that some major English scientists of that era wrote about perspective glasses is a testament to their popularity. Unfortunately, these lines of evidence did not appear designed to get everyone excited about new ways to admire the night sky, although the glasses might have been used for that purpose as Bacon suggests above (*New Atlantis*.) In his treatise *Rare Inventions* Michael Bourne describes how these

Fig. 5.4 Hariot's full map of the Moon. Image courtesy of Dr. Ewen Whittaker, and used with his kind permission

glasses work in words that are really describing a full two-lensed telescope: "For to see any small thing a great distance from you, it requireth the ayde of two glasses, and one glass must be made of purpose, and it maybe made in such fort, that you may see a small thing a great distance of, as this, to reade a letter that is set open neare a quarter of a mile from you, and also to see a man four or five miles from you, or view a Towne or Sastell, or to see any windowe or such like thing size or seauen myles from you" (Bourne 92).

Did Shakespeare actually use perspective glasses? If he did not, he certainly understood their operation well enough to admit them to *Richard II*, composed prior to 1597:

> For sorrow's eye, glazed with blinding tears,
> Divides one thing entire to many objects.
> Like perspectives which, rightly gazed upon,
> Show nothing but confusion—eyed awry,
> Distinguish form. (2.2.16–20)

Shakespeare's "perspective glass" was designed so that when an observer looks at it directly, the objects behind are blurred and indistinct; but by looking at a sharp angle, a sort of virtual eyepiece, the view comes into focus. I have experimented

Fig. 5.5 This replica of Galileo's telescope rests atop the author's 16-inch diameter reflector, with which he has discovered eight comets. The telescopes point towards a dome which houses an older telescope. Photograph by the author

with these views using one of my own telescopes. The glass in this simile, however, functions metaphorically not as a reflector of objects but as a reader of emotions. When looked at directly ("rightly gazed"), Bushy sees the Queen's mind as a jumble of conflicts, but when studied at an angle ("awry") her all-consuming dejection at the King's absence becomes clear. Around 1599, the perspective idea enters the canon again in a brief metaphoric remark in Henry V by the French King who exposes his virgin daughter Kate to a city whose walls have never been breached by an invading army:

> Yes, my lord, you see them perspectively, the cities
> Turned into a maid, for they are all girdled with
> Maiden walls that war hath never ent'red. (5.2.213–316)

The French King sees the cities as though they were a maid like his daughter. Whether any perspective glasses were used as telescopes, however, is another matter. Although no direct evidence exists that they were, perspectives probably were used as telescopes at some point; these lenses and mirrors made possible the magnification of objects over large distances. To write about glasses was nothing new in England at the turn-of-the-seventeenth-century. Lenses of many types populated the land. "The best kinde of glasse for this purpose," wrote Digges in Pantometria (1571).

> is of steele finely polished … neither concave nor convex, but flat and plaine as may be possible. This glasse it behoueth yee to hange up aboue the toppe of the cliffe with the polished side downwarde equidistant to the horizon wherein you must be great distance….

Fig. 5.6 The author stands next to some of his telescopes, *left* to *right*: Obadiah, a 12-inch (16-inch primary mirror) Schjmidt camera with a 7-inch apochromatic refractor on top; Pegasus, the 8-inch reflector he has used since 1964; Samwise, a 10-inch reflector, and Clyde, a Celestron 14-inch telescope. Photograph by Wendee Wallach-Levy

> This done, let a blumbeline fall from the centre of your glasse to the … ground platte on the toppe of the cliffe: (which ground platte) also you must with some diligence in the chayre thereof that it be … leuell and playne … your Glasse thus situate, turning your face towarde the shippe or other marke on the sea … go backwarde, always having your eye fixed on the glasse til such time as ye can see the shippe, or rather the very hull next to the water therein … (Digges 27).

Although Digges was not describing a telescope, he was writing about a device that does essentially what the telescope does, which is to gather light from a distance. All that remained, after the image was formed by the perspective glass or mirror, was to add a second lens to magnify the image. And this task was devised not just by the children of Hans Lippershey, but by several people working over many years from different lands. In his summary of Digges's work, Bourne writes of "the hollowe or concave glasse: and also that (convex) glasse, as ground and polished rounde, and thickest on the myddle, and thinnest towards the sides or eges" (Bourne 177). Although there is no written evidence that Digges or his contemporaries used these glasses to look at the sky, nor could I find any in my investigation of English literature of the time, there is some circumstantial substantiation of the possibility that they did.

According to the Canadian astronomer Helen Sawyer Hogg, Digges writes of "late Observations, to ratify and confirme hys (Copernicus's) Theories and

Fig. 5.7 What would Shakespeare have had to write about a telescope in space? In this 1997 NASA photograph, the wonderful Hubble Space Telescope is shown orbiting the Earth

Hypothesis. After half a century of research, the nature of these 'late observations' remains unclear, but it is probable that some of these were astronomical." Considering that these observations were designed to study the motions of the Moon, planets, and stars in order to demonstrate the efficacy of the Copernican model, it seems logical to assume that at least some of them involved pointing the perspective glasses toward the sky. We have already shown that this period was extraordinarily rich in celestial offerings. Several comets, particularly the Great Comet of 1577, would have lent themselves to study via perspective glasses (Hogg 201).

The name of Thomas Hariot appears frequently among those who are credited with being part of the invention of telescopes. His only published work, *A Briefe and true report of the new found land of Virginia* describes the native custom of "staring up into the heavens, uttering therewithal and chattering strange words and noises." To assist in their use, Hariot demonstrated a "perspective glasse whereby was shewed manie strange sights" (Hariot 22, 27). Hariot also mentions "the Eclipse of the Sunne which we saw the same year before in our voyage thitherward, which unto them appere very terrible" (27). Hariot probably was referring either to the partial solar eclipse of 10 May 1584, or to the eclipse of 29 April 1585. To Hariot, this was a natural phenomenon that he could have watched with the help of his perspective glass. Hariot also notes the passage "of a Comet which beganne to appear a few daies before the beginning of the said sicknesse" (Hariot 23). Hariot was probably referring to the comet of 1587, which was bright enough to cast a shadow (Yeomans 416).

If Hariot offers any hints that he used his perspective glass to look skyward, we must find them either in this one published work or in unpublished manuscripts such as the one Allan Chapman (322) was able to retrieve. Hariot left actual drawings of sunspots dominating the face of the Sun, and detailed representations of the Moon that far exceed Galileo's in terms of quality. Considering that these observations were conducted about the same time as those of Galileo, I support Chapman's suggestion that Hariot be given joint credit with Galileo for discovering the sunspots with a telescope. Chapman also lists Hariot's "first light" of his telescope the first night he used it on the sky as 26 July 1609, some weeks before Galileo tried the same thing. Hariott did not use any of his six telescopes to see Jupiter's moons before 14 December 1610, almost a year after Galileo. Although in his book Hariot comes tantalizingly close to pointing a perspective glass to the sky, either to view the eclipse or the illness-carrying comet, he ultimately failed to publish any observing session using his perspective glass. We can, however, imagine a sort of "English Renaissance Astronomy Club," a loose association of colleagues and friends that included Hariot, Thomas Digges, John Dee, and by extension Shakespeare, who gains admission after writing in *Hamlet* that

> Last night of all,
> When yon same star that's westward from the pole
> Hath made its course to illumine that part of heaven
> Where now it burns. Marcellus and myself,
> The bell then beating one,—(*Ham*. 1.1.39–43).

This circumstantial evidence comes from Bernardo describing the appearance of a great star westward from the pole. As Olsen (68) points out, strictly speaking, all stars are south of the pole but it is easy to place Cassiopeia towards the western horizon from Polaris, especially in mid-winter, sometime after King Hamlet was killed while napping in his garden on a presumably mild autumn afternoon. Besides Capella, there are no brilliant stars in that region of the sky; however the supernova of 1572 was visible in its position near Kappa Cassiopeiae for as much as 2 years after its sudden appearance in November 1572.

One cannot conclude from the material presented in this chapter that telescopes were invented in England, but a great deal of evidence shows us that the recognized early form of telescope called perspective glass was used in England by Leonard and Thomas Digges, John Dee, Thomas Hariot, and probably many others prior to the elder Digges's death around 1574. The literature of the time dealt with this technology in a number of ways. Shakespeare refers directly to them, and other writers, like Nashe, allude to the beauty of their effects; a mere line evokes the great beauty lying in a single piece of cathedral glass

> … which like the sun, that twixt two glasses plays,
> From one to th'other casts rebounding rays. (*Choice of Valentines* 200–202)

Nashe has successfully captured the optical bending of light that takes place, perhaps anticipating Newton's discovery of refraction of light by a prism. Usher proposes that "the front of Jove himself" represents "an eye like Mars, to threaten and command" (Hamlet 3.4.56–57). Usher takes the idea even further, proposing that on the face of Jupiter is an "eye like Mars" (Usher 1258); he defines the "eye" as Jupiter's Great Red Spot, a feature first published by Robert Hooke around 1664. Even if telescopes were in general use by 1620, I doubt they would have been good enough to allow an observation of Jupiter's red spot. To claim it discovery as early as 1600 strains credulity; Usher offers this as a portion of his "cosmic allegory" that he sees taking place in *Hamlet*. At a more basic level, particularly in Bernardo's identification of the bright star "westward from the pole" as being Tycho's star, Usher indeed offers the valid point that at a particular time of year, the supernova of 1572 would qualify as the star being pointed out.

The Telescope in 1610

The children of the Dutch spectacle-maker, Hans Lippershey, allegedly invented the first telescope as they played with their father's lenses. Upon learning of this new device in 1609, Galileo Galilei immediately assembled one with a magnification of ten and used it to discover craters and mountain ranges on the Moon in 1609. Chapman correctly points out that Hariot drew maps of lunar features some weeks, or even months, prior to Galileo (Chapman 27–33). In 1610, Galileo followed this up with his discoveries of the moons of Jupiter, spots on the Sun, and the phases of Venus. The telescope differed substantially from the older perspective glasses. Instead of a single slab of glass with a "virtual" eyepiece, the telescope introduced a second lens that users employed as a physical eyepiece. At the close of 2007, the United Nations voted overwhelmingly to celebrate this event: The quadricentennial of Galileo's use of the telescope in the year 2009 became the International Year of Astronomy (UN resolution 61/185, 20 December 2007).

Apparently no record survives to tell us whether Digges used his perspective glass to study the Milky Way, but it is surely possible that he did. Galileo was doubtless the first person to publish observations using a telescope with an objective lens and an eye lens specifically for astronomical observation and discovery; he recorded and published his discoveries before a wide readership, and he defended (unsuccessfully prior to 1642) the conclusions he derived from his observations before religious authorities. Galileo's published interpretations of these finds as they supported the Copernican theory resulted in his being shown, in 1633, the instruments of torture, tried for heresy, and forced to renounce his conclusions.

Galileo's discoveries galvanized the world, and double-lensed telescopes exploded in popularity. He published his discoveries in clearly understood Italian. Galileo's discoveries helped make the early modern period a time rocked by change not only in English society, but also in the tools that society used to interpret its literature. Galileo discovered in its orbit of the Sun, that Jupiter was accompanied by four moons, the first ever seen other than our own. The moons did not refute Ptolemy's cosmology, but their orbits about Jupiter were easily consistent with a solar system with the Sun at its center. During the fall of 1610, Galileo also made an extensive series of observations of Venus; although that planet had no moons, it did display phases which could be explained only by the Copernican theory.

Shakespeare and the Telescope

In his closing years Shakespeare did not ignore the early telescopic finds. *Cymbeline'*s grand ending scene shows the god Jupiter descending to the stage, surrounded by four angels; according to Muisano's "Shakespeare's Last Act" the scene can be interpreted as a representation of the planet Jupiter accompanied by the four moons discovered that year by Galileo. A difficulty with this reading is that the apparitions do not accompany Jupiter so much as they precede him, but it can be resolved by invoking Galileo's sketch of 13 January 1610, which does show all four moons westward (or preceding) Jupiter as it moves diurnally across the sky (Drake 52).

Had Shakespeare's prime writing years ended a decade later (after 1600), I wrote in 1995, his plays might have reflected a vastly different situation (Levy, 67).

After Galileo's crucial observations of Jupiter, Venus, the Moon, and the Sun, philosophers and writers contended with the mounting evidence for the new philosophy of a universe in which the Earth circles the Sun rather than one in which the Sun orbits the Earth. I was wrong; it turns out that Shakespeare's prime writing years ended in plenty of time to allude to the telescope. Besides *Cymbeline*, Shakespeare's great romances *Pericles* (1608), *The Winter's Tale* (probably 1609), and *The Tempest* and *Henry VIII* (1610) all were produced well within the time that they could have been affected by Galileo's discoveries. *Cymbeline*, especially, had appropriate timing, having been completed around the end of 1610.

References to the Sun-centered universe in writing before 1610 are rare, but after that critical year they become more common. (See Donne's *Second Anniversary* and Milton's *Paradise Lost* 5.261–263). Maisano offers good evidence that certain lines in *Cymbeline* hark back directly to Galileo's words in Starry Messenger. Maisano quotes Princess Imogen's lines in her Somnium:

> O learned indeed were that astronomer
> That knew the stars as I his characters;
> He'd lay the future open. (3.2.37–39)

Traditionally, "that astronomer" is interpreted as a judicial astrologer, but those particular lines could be read differently. Could Imogen be looking at someone who is simply watching the sky? The future he lays open is not an astrological prediction. Instead Imogen imagines a watcher of the sky who, like Galileo, "knew the stars." If this was Shakespeare's intention, then it may be his reaction to Galileo, whose discoveries of four moons of Jupiter earlier that same year appear as the ghosts preceding Jupiter in Posthumus's dream. Four apparitions—Posthumus's father Leonatus, his mother, and his two brothers—descend and surround him as he sleeps.

As the dream goes forward, Jupiter himself descends and the "ghosts" then "fall on their knees." Paisano (404–405) links the four apparitions that precede Jupiter to the four moons that Galileo has discovered using his telescope, but Jupiter's actual descent is accompanied by an indefinite number of ghosts. Either Shakespeare chose four apparitions by extraordinary coincidence, or, far more likely, the ghosts represent Jupiter's Galilean moons.

In Maisano's vision, the four apparitions appear at the start of Posthumus's dream. Jupiter does descend accompanied by an eagle and an undetermined number of additional ghosts. Could these ghosts represent still undiscovered moons? In either case it is exciting to find Shakespeare possibly responding to Galileo in this positive way. Jupiter lays a tablet on the sleeping Posthumus, who awakes soon after to find that it is actually a book, which Maisano (406, Usher 7–12) suggests, is Galileo's seminal *Siderius Nuncius*. Similarly at the end of *The Tempest* Miranda speaks of new ideas and new philosophies:

> O wonder!
> How many goodly creatures are there here!
> How beauteous mankind is! Oh brave new world
> That hath such people in't! (*Tempest* 5.1.183–186)

Usher suggests that the terms used in *Cymbeline*, "book" and "label", are "probably mundane references to *Siderius Nuncius*." Although Usher does not provide real evidence to support this claim, Shakespeare may well have thought of Jupiter's moons when he conceived the idea of Posthumus's four-member ghostly family. Furthermore, since *Siderius Nuncius* was an immensely popular book that year, it is a reasonable guess as to what Shakespeare intended. Considering the author's strong interest in the night sky as shown in so many other places, it is easier to conclude that these are references to Galileo's discoveries rather than to assume he ignored them. Moreover, the musical pomp and circumstance surrounding that scene lends credence to Galileo's idea of the dawn of a new age.

If Shakespeare did allude to Galileo in this late play, he may well have extended that insinuation in an even later drama, *The Two Noble Kinsmen*, which he allegedly coauthored with John Fletcher around 1613–1614. It was first performed at Blackfriars in the autumn of 1613 (Bertram 13), but Bertram uses Thomas Digges's son Leonard's claim that "Shakespeare did not collaborate" (Bertram 244) as evidence that Shakespeare was the sole author.

Invocations of Jupiter persist in this story. There are five allusions to Jupiter: First, the king of the Roman gods, or the "ruler of planets" acknowledges the honor of the bridegroom's bed (1.1.29). Second, later in the same scene the First Queen accuses Jupiter of destroying the effects of his actions "before they touch" (1.1.136). Third, the same monarch successfully prevents Jove from a synod or meeting (1.1.177–178). Fourth, after the battle scene in Act 3, Theseus swears by Jupiter's son Castor, that both kinsmen shall die (3.6.135). Fifth, in response, Palomon calls Emilia a "bright star" by Jupiter, which "followed" to the east of the bright star Spica in the predawn sky throughout the first few months of 1614 (3.6.146).

In her soliloquy Emilia, while carrying pictures of both kinsmen, recalls Jupiter's snatching Ganymede from Earth and forming "a shining constellation" with his new cupbearer (4.2.15–18). Simon Marius, who claimed discovery of Jupiter's brightest moons, had proposed the name Ganymede but it did not catch on by the time *Cymbeline* was performed (Mundus Iovialis, 78f). Jupiter and its moons did form a faint and moving asterism, telescopically and poetically a "shining constellation." Since Shakespeare is credited with the composition of all these scenes but 4.2, the multiple references to Jupiter could reflect on his earlier calling on this world in

Cymbeline (Oxford, 500). In his *Pleasant Dialogues* … Heywood offers a discussion between Jupiter as a mythological god and the servant he snatched while disguised as an eagle. Ganymede wants Jupiter to return him to Earth.

In *Kinsmen*'s fifth act, it is Venus, not Jupiter that plays a decisively judgmental role. Palomon is warned by Perithous:

> Noble Palamon, he warns, The gods will show their glory in a life, That thou art yet to leade. (5.4.57)

Palomon replies: Can that be, When Venus, I have said, is false (5.4.57–60).

Palomon probably meant that Venus was false not only in the obvious sense of disloyal but in a judicial astrological sense of being retrograde (Ganent 435). However I propose a different interpretation based on Venus's orbital motion. Galileo remarked in *Siderius Nuncius* that "as it (Venus) reaches its maximum departure from the Sun it be semicircular. From "full Venus", the planet wanes to a quarter phase and thence into a horned shape, once more becoming thinner as it approaches the sun" (Drake 93). A few lines later, Paisano adds:

> …to the goddesse Venus Commend we our proceeding, and implore Her power unto our partie. (Here they kneele as formerly.) Haile, Soveraigne Queene of secrets, who hast power
> To call the feircest Tyrant from his rage,…
> and induce Stale gravitie to daunce … (5.4.89–109).

A simple reading of *Siderius Nuncius* shows how, in the early fall of 1610, Galileo began his visual studies of Venus, a planet then going through a poor apparition in the southwestern sky in the early evenings. (Drake, 93–94) Taking advantage of Venus's presence from the end of August 1610 to early February 1611, and hoping to discover moons like Jupiter's, Galileo was astonished to find something very different. There were no moons. Instead, over several months, Galileo observed the planet's appearance change radically from a full sphere, to an awkward gibbous then half-world, and finally to a crescent. Venus, Galileo concluded, has phases like Earth's moon. Galileo believed that although it was possible to explain Jupiter's moons in an Earth-centered system, it was not possible to explain Venus's phases in any other way than to conclude that Venus orbits the Sun in a closer orbit than the Earth.

Venus is indeed a "queen of secrets." If *The Two Noble Kinsmen* was really composed around 1613 (and most scholars who accept the Shakespeare-Fletcher collaboration ascribe the writing of Act 5 scene 4 to Shakespeare), then Venus may have played as big a role as a planet as much as an astrological symbol. Venus's judicial astrology is powerful enough to induce dancing by dint of its own gravity, or to calm a tyrant's rage. But in a modern way it symbolized the greatest of Galileo's discoveries, that Venus shows phases. Because it provided solid evidence to discredit the Ptolemaic system, the discovery of the phases of Venus was really Galileo's most important observation.

Because Shakespeare's concluding play was completed 3 years before his death, his collaboration with Fletcher to complete it was accepted by many critics and more important, Shakespeare might have been using Galileo's accomplishment as an image, perhaps even dipping into Yeats's *Spiritus Mundi*, as I proposed back in this book's preface, preface, in his play. The wedding scene invokes Mars, a conventional male allusion though Mars the planet was also an ideal subject for telescopic observation during its apparition of September 1610. However, the emphasis on Venus does invite this cosmological interpretation. "Before the alters of Mars and Venus, it is clear that it is two kingdoms for which Arcite and Palomon contend" (Bertram 278). As Usher is so insistent on a cosmic allegory for *Hamlet*, Mars and Venus imply a more comprehensible extension of the metaphor into this later play.

Allusions to the telescope appear in other works of the period, notably Beaumont and Fletcher's *A King, and No King* wherein one citizen thinks he sees the servant Philip in the distance; the other says "Thou wouldst be Philip, if thou sawst it in a glass; it looks so like a Visour." (2.1.100)

We have already in Chap. 4 that Webster's *The Duchess of Malfi* deals extensively with the astrology of its time. The play also alludes to Galileo and his telescope in one of the Cardinal's speeches:

> We had need go borrow that fantastic glass
> Invented by Galileo the Florentine,
> To view another spacious world i'th' moon
> And look to find a constant woman there. (*Duchess*.2.4.25–28)

The image suggests that we cannot find a constant woman anywhere on Earth, and that we must borrow Galileo's telescope to look for one on the Moon.

In his *Albumazar*, Tomkis refers directly to Galileo's work in the title character's speech:

> Ronca, the bunch of planets new found out
> Hanging at the left of my best perspicill,
> Send them to Galileo at Padua;
> Let him bestow them where he please. But the stars
> Lately discovered twixt the horns of *Aries*
> Are as a present for Pandolfoe's marriage … (1.5.1–7)

However, Albumazar cannot be referring to moons of Jupiter as "the stars lately discovered" since Jupiter was in Taurus, not Aries, in January 1610. The planet and its moons would have been in Aries a year earlier.

Nevertheless, these many references illustrate a symbiosis that was forming between the explosively developing science of the time, and the English Literature that accompanied it. Galileo, silenced and sentenced to spend the remainder of his long life under house arrest, became popular as an English literary figure (Figs. 5.1, 5.2, 5.3, 5.4, 5.5, 5.6, and 5.7). Ultimately, he appears to be the *only* person cited by name in *Paradise Lost*:

> Through Optic Glass the Tuscan artist views
> At Ev'ning from the top of Fesol (PL. 1.288–289)
> There lands the fiend, a spot like which perhaps
> Astronomer in the sun's lucent orb
> Through his glazed optic tube yet never saw (3.588–590)
> … and finally, by name
> As when by night the Glass
> Of Galileo, less assur'd, observes
> Imagind Lands and regions in the Moon: (*PL* 5. 216–263)

Part II
The Sky in the Poetry of Tennyson and Hopkins

Chapter 6
Bridging the Centuries: Astronomical Discoveries Between the Eras of Shakespeare and of Tennyson and Hopkins

"Nova stella, novus rex!"
-Bayeux Tapestry inscription above Halley's Comet

The time of Shakespeare's life and the two centuries that followed his era and preceded the time of Tennyson and Hopkins were filled with advances. Shakespeare was well aware of the early uses of the telescope, and possibly alluded to Galileo at the end of Cymbeline, but much more must be covered to fill the gap between the celestial knowledge of the early 1600s and of the mid 1800s. This chapter will briefly overview the major figures and findings.

Tycho's Star

During Shakespeare's own lifetime, Tycho Brahe lived at Uraniborg and made, along with several others, one of the most significant discoveries in the history of science. The reason that the supernova is credited to him as "Tycho's star" is that he observed the star intensely after discovery and he write a treatise about it.

A large nobleman with a long red moustache and beard, and the flair of a swashbuckler, Tycho Brahe is one of the earliest examples of an astronomical discoverer. Some finds take years of careful research. Tycho's first discovery took 2 seconds, and they had an indelible effect on the course of history. Born December 14, 1546, Tycho became one of the greatest astronomers who ever lived. He is credited with settling the observational groundwork that led to the confirmation that the Earth is not in the center of the Universe. Though he had the greatest observatory of his time on Earth, he preferred to be thought of as an aristocrat more than an astronomer. He allegedly always wore full court regalia while observing because, most likely, that lent distinction to his pursuit. Science in the 1500s was looked down upon, and so

D.H. Levy, *The Starlight Night*, Astrophysics and Space Science Library 419,
DOI 10.1007/978-3-319-19878-1_6

were those who philosophized about it and studied it. Tycho did not want to let the fact that he practiced science interfere with his lofty reputation as a nobleman.

Early in his life, Tycho was a cocky student in the German city of Rostok. There, the big man was partying at a pre-Christmas dance when he got into an argument with another party guest, Manderup Parsbjerg. The dispute was broken up, but at a Christmas party a few nights later they argued again, this time much more loudly. They decided to settle their dispute in a duel two nights later. Swords clashed. Although history doesn't confirm what happened to Parsbjerg, Tycho did not fare well. The tip of Parsbjerg's sword found its way into Tycho's nose, slicing off its bridge. Tycho was able to stop the bleeding, but the resulting scar was unsightly and undignified. He spent considerable time, and took advantage of his own metal casting abilities, to fashion a prosthesis of some skin-colored copper alloy that was attached to the end of his nose. Tycho turned what could have been a major embarrassment into a sort of mark of honor. (Ashbrook, Tycho Brahe's Nose, 3–5.)

Tycho did not learn to control his temper in the years after the duel. He lived for a time with his father, who died in 1571, and then with his uncle, where he developed his interests in chemistry and alchemy. Tycho was letting his astronomical interest lapse. This respite ended suddenly on the clear evening of November 11, 1572. He was stunned beyond words to see a bright new star adding to the five bright members of Cassiopeia, the constellation of the ancient Queen of Ethiopeia. He did not believe the evidence of his own eyes, and asked his servants and local farmers to report what they saw as well. When he got over his shock he began making detailed measurements of its position and brightness. His own measurements indicated that the supernova was in Cassiopeia at 0 hours, 22 minutes, and +63 degrees 53 minutes. (*Sky & Telescope*, October 1944, 12; position updated to 1950 coordinates.) The remaining nebula is at least 22 magnitudes fainter than the star was at its maximum.

Tycho's measurements of this bright *stella nova* confirmed that it did not move among the distant stars, and therefore must belong to that outermost sphere, or area of space, that lay beyond the Moon. Visible in broad daylight at its maximum magnitude of about minus 4 (as bright as Venus) November 11, 1572, the star faded in brightness and disappeared in 1574. But the star also changed color as it faded. At the time, most scientists still accepted Ptolemy's Earth-centered Universe theory, which also held that the eighth sphere, which contained the fixed stars, was supposed to be unchanging. By far Tycho's most important conclusion was that the star proved that the distant heavens are not, as had previously been thought, immutable.

Shortly after the discovery of the new star, Tycho moved with his wife and growing family to Copenhagen. The place where he made most of his contributions to astronomy, however, was at his island observatory on Hven, some 15 miles northeast of Copenhagen. Tycho was unquestionably lord over the entire island of Hven, an honor bestowed on him by none other than Denmark's King Frederick II, a man with a great interest in the sciences. (No longer a part of Denmark, Hven became a part of Sweden more than half a century after Tycho's death. Today Hven remains a rural island, with a single village, Backviken, on its southeastern shore.)

On this island, Tycho lived in his home and observatory, called Uraniborg, the site of major observations from its completion in 1576 to the time he was forced to abandon it in 1597. Never forgetting that he was a nobleman, he designed Uraniborg with the attributes of a castle. The main Uraniborg building was 49 feet square and some 37 feet in height. Reflecting the cosmic vision of its builder, the building's four front doors faced the four cardinal compass points. Uraniborg offered some highly advanced features for a sixteenth century establishment, including running tap water and central heating. The observatory included two features almost unheard of before or since. One was an author's dream of a paper mill and printing press, so that Tycho could publish his books and manuscripts independently, and the other was a prison.

Galileo and Milton

Galileo's accomplishments and tragedy deserve many words. He is a major and noble figure in the history of science. For the purpose of this book, I will focus on one single event in his life. During the summer of 1638 a young John Milton visited Galileo. At least one critic, G.F. Butler, counters that Milton never met the great scientist, because there is no confirming evidence of it. (Butler, 132–139) But unless Milton lied in *Areopagitica*, or meant the piece to be fiction, in Areopagitica Milton wrote clearly:

> There it was that I found and visited the famous Galileo, grown old, a prisoner to the Inquisition, for thinking in astronomy otherwise than the Franciscan and Dominican licensers thought. And though I knew that England then was groaning loudest under the prelatical yoke, nevertheless I took it as a pledge of future happiness, that other nations were so persuaded of her liberty. (Milton, 487)

While arguments persist as to whether Milton really did visit the iconic scientist, his point stands in using Galileo as an example of what could happen under the terror of censorship. The question is as irrelevant as asking whether James Joyce had the slightest evidence that Shakespeare actually took his boyhood summer walk in 1577, more than three centuries before *Ulysses* was written. In both cases the authors point out that these events could have taken place. On the same level, one could argue that Albert Einstein never played the violin, or that I never met Clyde Tombaugh in 1963, Bart Bok in 1979, or Gene and Carolyn Shoemaker in 1988. Ultimately, Galileo appears to be the *only* person cited by name in *Paradise Lost*:

> Through Optic Glass the Tuscan artist views
> At Ev'ning from the top of Fesol (PL. 1.288–289)
> …
> There lands the fiend, a spot like which perhaps
> Astronomer in the sun's lucent orb
> Through his glazed optic tube yet never saw (3.588–590).
> …
> … and finally, by name
> As when by night the Glass

Of Galileo, less assur'd, observes
Imagind Lands and regions in the Moon: (5. 216–263)

Huygens

In 1655, at the young age of 26, Christiaan Huygens used a telescope with a focal length of 12 feet with which he discovered, in March 1655, Saturn's largest moon. Now named Titan, this moon, almost as big as Mars, is the largest in the solar system. It has a dense atmosphere. In 2004, a probe named Huygens after Titan's discoverer, sailed into Titan's atmosphere and explored the surface of this mysterious world.

Huygens also used this telescope to try to resolve a major enigma concerning Saturn that had lasted almost half a century. The problem began when Galileo, in July, 1610, turned his telescope to Saturn, which had recently become visible after several months of invisibility due to its being too close in the sky to the Sun. He wrote: "For the same reason I have resolved not to put anything around Saturn except what I have already observed and revealed—that is, two small stars which touch it, one to the east and one to the west, in which no alteration has ever yet been seen to take place and in which none is to be expected in the future…" (Galileo, 101) Through his telescope he saw two ear-like appendages that, over several years, grew smaller and vanished. Then after a season in which the appendages were completely absent, they returned—small at first, then larger.

Huygens was only 26 when he peered through his new telescope at Saturn. He did not see any appendages, but instead, the view through his telescope hinted strongly that Saturn was surrounded by a ring. Reluctant to announce the discovery at first, he translated his message into a simple Latin sentence, and published it as an anagram:

AAAAAA CCCCC D EEEEE G H IIIIIII LLLL MM NNNNNNNNN OOOO PP Q RR S TTTTT UUUUU. When worked out, the anagram read:

Annulo cingitur, tennui, plano, nusquam coherente, ad eclipticam inclinato. Translated to English, it meant:
It is surrounded by a thin, flat ring, nowhere touching, inclined to the ecliptic.

Comets

A single remarkable comet, returning over and over again, has punctuated human history. Although the English astronomer Edmond Halley determined by the middle 1700s that the comets of 1531, 1607, and 1682 were actually separate and repeat visits by the same object, this comet has been playing on humanity's interests and fears for centuries. In a fortuitous series of coincidences, Halley's Comet appeared at such critical moments as the defeat of Attila the Hun in 451 AD, and the Norman conquest of England in 1066. It was that latter visit that underscored the perception that the heavens themselves still blazed forth the death of princes. "*Nova stella, novus rex!*" ("New star, new king!") was the battle cry on the beautiful tapestry at Bayeux that later wove the story of that battle. By the fifteenth century that maxim still had not changed. A story from *The Illustrated London News* tells the story of a comet, not that of Halley, whose untimely visit caused the death of a prince. The guilty comet appeared in 1402 and was visible in broad daylight for a week:

> There is no doubt, however, that comets sometimes really did produce fatal effects. In June, 1402, one appeared in Italy which literally killed the famous John Galeas Visconti. The astrologers of the Prince had predicted that his death would be announced by a comet of extraordinary magnitude, and the celestial phenomenon had no sooner become visible than his Highness, speechless from fright, sank to the ground and died. (Illustrated London News, August 1861.)

One might suspect that the terror struck into the human mind by omens such as comets and exploding suns caused his Highness's death, and that such coincidence no longer belongs to our history. Yet it does. As recently as March 1997, 39 members of the cult called Heaven's Gate committed mass suicide to follow their belief that a UFO would usher them to a space vessel travelling behind Comet Hale-Bopp, which was rounding the Sun at that time.

When the breakthrough finally came that allowed us to see comets as full-fledged members of the solar system, it had nothing to do with their appearance or structure. It had everything to do with their orbits, or paths around the Sun, and with the genius of one man, Edmond Halley. Born November 8, 1656, this son of a soapmaker and salter had a keen interest in science. As a 20-year-old student at Oxford he published his first paper. He wasn't one to stand on ceremony, however; impatient with Oxford, he dropped out and headed south that same year to the island of St. Helena, the island that, more than a century later, would serve as Napoleon's exile home after the Battle of Waterloo.

Young Halley's mission to that south Atlantic island came about because he realized that the southern sky was virtually unexplored. From his visit he charted many objects that were completely invisible from England. Halley's observations resulted in *Catalogus Stellarum Australium*, a catalog of stars in the southern hemisphere that the young astronomer published in 1678. Halley's mentors at Oxford were so delighted with the catalog that they allowed him to re-enter Oxford, and later awarded him a master's degree without requiring him to take the exams. Coupled with observations made later in his long life, Halley would discover that stars do change their positions relative to each other. He also inspired expeditions around the world to see Venus in two of its rare transits, as the planet passed in front of the Sun in 1761 and 1769, arranged in response to his suggestion that observations of these transits could allow a precise calculation of the distance between Earth and the Sun. We remembered Halley again during the transits of Venus in 2004 and 2012.

Halley then turned his attention to comets, the subject that would ensure his legacy and fame, those objects that move through the sky at supposedly irregular intervals. Thinking that the most productive way to study comets was to explore their orbits, Halley decided to study the orbits of as many comets as possible. His successful completion of this task inspired him to work out the orbits of 21 other comets that appeared from 1698 all the way back to 1337. Three of these comets had almost identical orbits. He quickly noted also that these comets appeared at intervals of roughly 76 years, in 1531, 1607, and 1682.

Although Halley suspected that the comets of 1378 and 1456 were earlier appearances of the same comet, he was not certain of this. There was another in 1301, but it was 4 years too early to be a return of this particular comet. It was: the comet's orbital period, it later turned out, was varying from 74 to 79 years, depending on perturbations from the planets. It was not until the nineteenth century that celestial mechanicians like J. Russell Hind were able to connect earlier apparitions to Halley's comet, and early this century Cowell and Crommelin used ancient records to confirm the comet's visit as far back as 240 B.C. Over the centuries, the effect of "planetary perturbations" would be measured to have varying effects on many comets. For some, the effect would be negligible. For others, like Halley, the effects were considerable and easily measurable. For Comet Shoemaker-Levy 9, the consequences were catastrophic, leading to the comet's complete destruction.

Back in the mid-1700s, Halley was uncertain just how severe the problem of planetary attractions on comets was. As he grew older, Halley couched his predictions in progressively less certain terms. In Halley's first prediction in 1705, notes Caltech's Donald Yeomans, Halley wrote in Latin that "I shall venture confidently to predict its return in 1758." When he translated that into English later that year, it appeared as a far more modest "I dare venture to foretell…." Ten years later Halley hedged even more with "I think, I may venture to foretell…;" and in his last writings he meekly said became "if… it should return again about the year 1758…." (William Sheehan, 132). One of today's most experienced orbit computers, Yeomans believed that as Halley grew older he became more aware of just how complicated the calculation of comet orbits really is when the effects of the planets are taken into account. After all, he could predict all he wanted, but his comet was out there beyond the orbit of Neptune, and only that ball of ice and dirt could know precisely where it was and when it would return. Comets are like cats, I say. They both have tails, and they both do precisely what they want to do.

Halley died in 1742, after a full life of service to astronomy. As we all know, Halley did not live long enough to see his forecast come true. During 1758 an veritable army of mathematicians and astronomers were frantically searching for the comet, on paper and in the sky. It was not until Christmas night 1758 that Johann Georg Palitzsch, a Dresden farmer, spotted the comet with a small telescope, and the French Astronomer Charles Messier confirmed its appearance in 1759. This achievement inspired the following generation to continue the mathematical search for other possibly returning comets. Six years later, in 1765, Nicolas_Louis de Lacaille first referred to the comet as Halley's Comet. (K. TG. Jones, 347). For more than half a century, it stood alone as the only comet proven to be periodic.

Probably the most inspiring words ever set to paper about Halley's Comet derive from Leslie Peltier's biography *Starlight Nights*. An amateur astronomer who gained fame for his discoveries of 12 comets and for the 132,000 variable star observations he completed, Peltier's life story appeared in 1965:

> Within historic times twenty-eight visits (now thirty) by Halley's Comet have been recorded. On an early trip it witnessed the defeat of Attila's Huns in A.D. 451. It arrived in time to preside over the Norman Conquest in 1066. In the year 1459 the menacing appearance of the comet so alarmed Pope Calixtus that he decreed several days of prayer and established the midday angelus. With a great clanging of bells he then besought the comet to visit its wrath solely on the invading Turks. In 1607 it was watched by both Shakespeare and Kepler and I like to think that it was also seen by Captain John Smith and Pocahontas in the frontier skies of Jamestown. On its following trip around in 1682 the comet was observed by Halley himself, who probed into its periodic past and bequeathed to it an honored name that it can bear with pride throughout the solar system. By 1835, when it returned, affairs of earth had speeded up. Many a canal boat traveler, looking down, could see the comet glowing on the surface of his highway. Man himself had taken to the skies when the comet last appeared in 1910, for he was making fledgling flights of perhaps one hundred miles. In 1986 (Peltier's book was published in 1965, well before the comet's 1986 appearance) our historic visitor will be visited in turn, for in that year a spacecraft from earth will hold a rendezvous with Halley's Comet pout in space! (The European Giotto spacecraft did make that meeting, sending to humanity its first images of the comet's potato-shaped nucleus.)

Who would venture to foretell the wonders and achievements which the comet will witness in that distant year of 2062? Or will man himself prove periodic? Will the Huns be back again? (Peltier, 16–17).

Charles Messier: The Comet Ferret

Comet Halley's visit in 1759, followed so assiduously by the famous French observer Charles Messier, was a watershed for comet observing. Other comets had been shown to be periodic; Anders Johan Lexell computed that a comet discovered by Charles Messier in 1770 had a period of only 5 years. But there was more: Lexell showed that it had come close to Jupiter just before its approach to the Earth, and that afterward it approached Jupiter a second time. In this later encounter, Jupiter's gravity thrust the comet right out of the solar system. There is also the work of Johann Franz Encke, who linked the apparitions of comets that appeared in 1786, 1795, 1805, and 1819, to fit the orbit of a single comet that would next return in 1822. Encke was right, and the comet which now bears his name has by far the shortest orbital period of any comet, a brief 3 1/3 years. (I made an independent discovery of Encke's Comet while comet hunting in 2000.)

I suspect that Messier got his comet hunting start thanks to his failure to be the first person to see Halley's Comet in 1758. Perhaps he wouldn't admit it, but I like to imagine that he decided to get even with the sky, so to speak, by finding every other new comet that appeared after he was beaten out by Johann Georg Palitzch, a farmer from Dresden, in the race to spot the most important comet of all. On January

21, 1759, while working at Nicholas Delisle's observatory at the Hôtel de Cluny in Paris, Messier finally saw the comet. He was thrilled by it—"It was one of the most important astronomical discoveries," he wrote, "for it showed that comets could return." (Jones, 347.) He summoned Delisle, who observed the comet, and then promptly ordered Messier not to announce it in any way. Delisle's clandestine behavior was not corrected and the comet announced until April 1, 3 weeks after the comet had already rounded the Sun. By this time, Messier had already heard of Palitzch's Christmas Night sighting and knew that he would have lost the race anyway, despite anything Delisle did.

Messier didn't have long to wait to assuage his disappointment. Discovering his first comet in 1760, he made his first eight finds with the playing field virtually to himself. He found comets in 1763, 1764, 1766, 1769, 1770, 1771, and 1773 before his chief rival Jacques Montaigne, a druggist in Limoges, France, snared his first in 1772, the comet that in 1826 became known as Biela's Periodic Comet or P/Biela. This comet apparently split in two in 1846, returned as two comets in 1852, and only as a meteor storm in 1872 and in 1877. By 1781 Pierre Méchain had also joined the contest. But in 1801, Messier was still in the lead, having added six more notches to his telescope.

More Important than a Comet

Friedrich Wilhelm Herschel would become one of the greatest figures in the history of astronomy, but in 1760 he was an impoverished musician who didn't have the funds to return to England from a trip to Genoa, Italy. Desperate to get home, he gave a bizarre impromptu recital—a maestro playing a harp, holding a horn, and having a second horn attached to his shoulder. It might be hard to picture Herschel playing all three instruments in such a pose, but the people of Genoa came in droves and he made his way back to England.

Born in Hanover, Germany, in 1738, Herschel was the son of a musician, and he became a composer and conductor as well as a skilled performer on horn, harp, and organ. He used the English version of his first name after he moved to England in 1757. Herschel wrote 24 symphonies, plus some concertos. Being a contemporary of Mozart, his music reflects the style of that master.

In May of 1773, Herschel made a small purchase that would have major consequences. He records in his diary: "May 10. Bought a book of astronomy and one of astronomical tables." (Lubbock, 60). Within a few weeks his passion for the order and harmony of music had transformed itself into that of the sky, and he plunged into his new love with his typical zeal. Buying some lenses, he built a four-foot long tube and mounted a lens in it: "With this," he exulted, "I began to look at the planets and the stars. It magnified 40 times. In the next place I attempted a 12 feet one and contrived a stand for it. After this I made a 15 feet and also a 30 feet refractor and observed with them."

By 1781, only 8 years after he fell in love with the stars, Herschel was deeply involved in a systematic survey of the sky. "On Tuesday the 13th March," he wrote, "between ten and eleven in the evening, while I was examining the small stars in the neighborhood of H Geminorum, I perceived one that appeared visibly larger than the rest; being struck with its uncommon magnitude I compared it to H Geminorum and the small star in the quartile between Auriga and Gemini, and finding it so much larger than either of them, suspected it to be a comet". (Dreyer,1: 30–38.) Although the object was pretty bright, perceiving it not as a point of light but a disk was a real challenge even to a trained eye. "Seeing is in some respects," Herschel wrote later, "an art which must be learnt. To make a person see with such a power is nearly the same as if I had been asked to make him play one of Handel's fugues upon the organ. Many a night have I been practising to see, and it would be strange if one did not acquire a certain dexterity by such constant practice." (Hoyt, 12)

News of this peculiar discovery spread swiftly. Across the English channel, the famous comet discoverer Charles Messier observed the new object at every opportunity. Messier was taken aback by both its slow movement from night to night and its shape. "I am constantly astonished at this comet," he wrote to Herschel in the late spring of 1781, "which has none of the distinctive characters of comets, as it does not resemble any one of those I have observed, whose number is eighteen …"

On August 31, 1781, Anders Lexell concluded that the orbit of this new object was almost circular, and it never gets closer to the Sun than 16 times the Earth's distance to the Sun. The pieces of the puzzle had finally fallen into place. Herschel's object did not look much like a comet because it was not a comet. It was a planet, the first to be discovered in historic times.

We have now reached William Herschel, whose life coincided with romantic poets like Keats. Shakespeare's time, with its reliance on what could have been viewed with the naked eye, saw few astronomical discoveries, but later centuries were times of increasingly numerous finds that changed humanity's basic understanding of the sky. Keats, for example, compared his own revelation of reading George Chapman's translation of Homer to

> …some watcher of the skies
>
> When a new planet swims into his ken; … (Chapman's Homer 9–10).

Composed in October 1816, this Petrarchan sonnet is canonically interpreted to invoke William Herschel's discovery of the planet Uranus in 1781, although I connect the reference instead to the discovery by Flaugergues of the Great Comet of 1811 (Kronk 27). Six decades later, to use a prose example, Mary Anne Evans (a.k.a. George Eliot)'s *Middlemarch* asked probing questions about the kind of person who would make an astronomical discovery:

> Does it seem incongruous to you that a Middlemarch surgeon should dream of himself as a discoverer? Most of us, indeed, know little of the great originators until they have been lifted up among the constellations and already rule our fates. But that Herschel, for example, who "broke the barriers of the heavens"—did he not once play a provincial church-

organ, and give music-lessons to stumbling pianists? Each of those Shining Ones had to walk on the earth among neighbors who perhaps thought much more of his gait and his garments than of anything which was to give him a title to everlasting fame: each of them had his little local personal history sprinkled with small temptations and sordid cares, which made the retarding friction of his course towards final companionship with the immortals. (Eliot 108–109)

Eliot's vivid prose harks back to the early modern period's obsession with astrology, lifting Herschel up to the constellations. It can be compared favorably with *Julius Caesar*, whose sky was "a tempest dropping fire" (*JC*.1.3.10) during the dark hours before Caesar was assassinated.

Herschel died in 1822, when Tennyson was 13 years old. Hopkins was born in 1844, at the height of the Victorian period. Other astronomers followed Herschel, like his son John, who helped chart the southern sky using his father's 20-foot reflector. Hopkins and Tennyson both absorbed the enthusiasm for astronomical discovery (Figs. 6.1, 6.2, and 6.3).

Fig. 6.1 This is Comet Levy (C/1990 K1) which I discovered on May 20, 1990, which crossed the Milky way during the Summer of that year, surrounded by many "blessed candles." Photograph by the author. This comet brightened the evening sky during the Summer of 1990. It is an example of a comet and might be similar to one that Hopkins might have observed. Photograph by the author

Fig. 6.2 An image of Comet Levy (P/2006 T1) taken the morning after its discovery. The comet is the faint object to the *right side* of the photograph; Saturn is the bright object at *left*. Even though this was a relatively bright object, it really does seem "scarce worth discovery" as Hopkins wrote in his poem. Photograph by the author

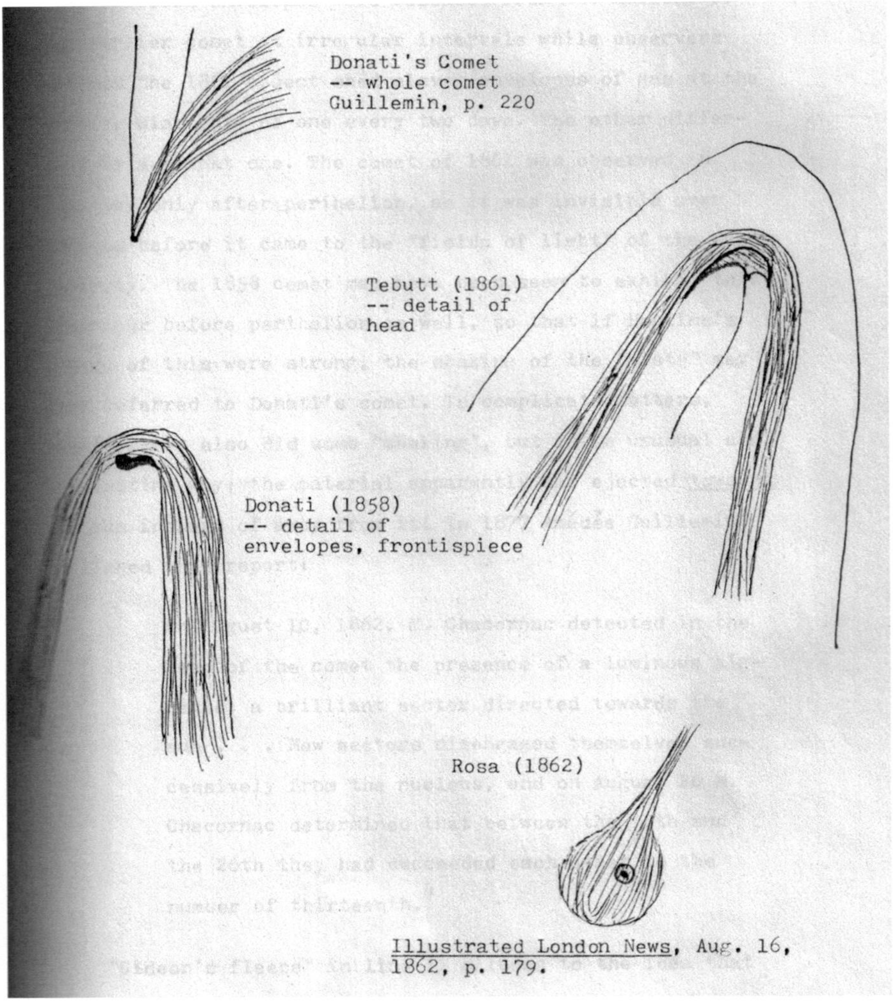

Fig. 6.3 Examples of Cocooning Mists. Prepared by the author with assistance from Dr. J. Ashbrook

Chapter 7
Leading up to Literature in the Victorian Age: Moving Toward Hopkins and Tennyson

This book evolved from the idea that literature and the night sky could, with certain authors, be studied together to their mutual benefit. To suggest lines on which such a study might proceed, it is necessary to choose a writer whose total output was not vast, and who had an intelligent interest in the night sky. The writings of Gerard Manley Hopkins and Alfred Lord Tennyson have provided most fertile ground for such a project. The knowledge that appears in his writings is underlain by a confidence in his ability to observe the night sky intelligently. Hopkins in particular used this knowledge throughout his life, even in his later retreats when he attempts to explain events described in Revelation by scientific means. Such confidence is rare in literature, but over the centuries there have been some authors who have carried into their writings an interest in the sky. Notwithstanding his thorough examination of a major scientific instrument in his Treatise on the Astrolabe, Chaucer makes fun of an astronomer in The Miller's Tale:

> He walked in the feeldes, for to prye
> Upon the sterres, what ther sholde bifalle,
> Til he was in a marle_pit yfalle:
> He sauah nat that.

The period around 1600, Shakespeare's time, was a critical one for astronomy. Even though Copernicus had published his ideas about a sun-centered universe in 1543, the general population, and poets, took little notice of them as these ideas were entirely theoretical. Shakespeare's lines from *Hamlet* suggest that the Bard himself did not take the new developments in astronomy very seriously:

> Doubt thou the stars are fire,
> Doubt that the sun doth move,
> Doubt truth to be a liar,
> But never doubt I love. (*Hamlet*.2.2.116–119)

D.H. Levy, *The Starlight Night*, Astrophysics and Space Science Library 419,
DOI 10.1007/978-3-319-19878-1_7

This general astronomical complacency was shattered in 1610 when Galileo first turned his telescope towards Jupiter and discovered four moons orbiting around it and not around the Earth. The reaction in poetry was swift. John Donne, in his *Anatomie of the World*, commented on the new philosophy only 1 year after Galileo's discovery:

And new philosophy calls all in doubt,
The Element of fire is quite put out:
The Sun is lost, and th'earth, and no mans wit
Can well direct him where to looke for it.
And freely men confesse that this world's spent,
When in the Planets, and the Firmament
They seek so many new: they see that this
Is crumbled out againe to his Atomies.
'Tis all in pieces, all cohaerence gone:
All just supply and all Relation:
Prince, Subject, Father, Sonne, are things forgot,
For every man alone thinkes he hath got
To be a Phoenix, and that then can bee
None of that kinde, of which he is, but hee.

The invention of the telescope in 1610 changed much more than our concept of the sky: it altered our perceived role in it. Donne, in this passage, is distressed by the loss of the old and familiar system and shows little confidence in the new order. And a decade later, in 1622, Sir John Davies shows contempt for it, since it interferes with the orderly cosmic dance he envisages in *Orchestra*:

Onely the Earth doth stand for ever still,
Her rocks remove not, nor her mountaines meete,
(Although some witts enricht with Learnings skill
Say heaven stands firme, and that the Earth doth fleete
And swiftly turneth underneath their feete)
Yet though the Earth is ever stedfast seene,
On her broad breast hath Dauncing ever been.

After the discovery of Uranus in 1781, telescopes became more popular in England. George Adams and his son made a series of beautiful Gregorian reflectors with metal mirrors mounted in brass tubes; in fact I owned one while writing the original of this part as my M.A. thesis. Meanwhile, other optical experts such as Jesse Ramsden crafted some small portable refractors that made stargazing easily accessible; I possessed one of his refractors as well. Wordsworth comments on the disappointment that he perceived in a crowd looking through such a telescope, in his 1806 poem "Star Gazers":

What crowd is this? what have we here! we must not pass it by;
A Telescope upon its frame, and pointed to the sky:
Long is it as a Barber's Poll, or Mast of little Boat,
Some little Pleasure-Skiff, that doth on Thames's waters float.

The Show-man chuses well his place, 'tis Leicester's busy Square;
And he's as happy in his night, for the heavens are blue and fair;
Calm, though impatient is the Crowd; Each is ready with the fee,
And envies him that's looking—what an insight must it be!

Yet, Show-man, where can lie the cause? Shall thy Implement have blame,
A Boaster, that when he is tried, fails, and is put to shame?
Or is it good as others are, and be their eyes in fault?
Their eyes, or minds? or, finally, is this resplendent Vault?

Is nothing of that radiant pomp so good as we have here?
Or gives a thing but small delight that never can be dear?
The silver Moon with all her Vales, and Hills of mightiest fame,
Do they betray us when they're seen? and are they but a name?

Or is it rather that Conceit rapacious is and strong,
And bounty never yields so much but it seems to do her wrong?
Or is it, that when human Souls a journey long have had,
And are returned into themselves, they cannot but be sad?

Or must we be constrain'd to think that these Spectators rude,
Poor in estate, of manners base, men of the multitude,
Have souls which never yet have ris'n, and therefore prostrate lie?
No, no, this cannot be—Men thirst for power and majesty!

Does, then, a deep and earnest thought the blissful mind employ
Of him who gazes, or has gazed? a grave and steady joy,
That doth reject all shew of pride, admits no outward sign,
Because not of this noisy world, but silent and divine!

Whatever be the cause, 'tis sure that they who pry and pore
Seem to meet with little gain, seem less happy than before:
One after one they take their turn, nor have I one espied
That doth not slackly go away, as if dissatisfied.

More than two centuries later, and using far better telescopes, I have noticed the same phenomenon. People who look through a telescope for the first time do not know what they are expected to see, and when they do not see things as clearly and in as much detail as shown by the Hubble Space Telescope, they still leave "dissatisfied". Only after they acquire some experience do they really appreciate what a real telescope has to offer. Wordsworth's comment probably could not be explained by the quality of early nineteenth century optics. The two telescopes I have just mentioned outperform even some of the small telescopes that are available today.

Just 5 years after "Star Gazers", in 1811, the French stargazer Jean-Louis Pons discovered a comet destined to be one of the best and brightest in history. He found it 3 weeks after the French observer Honoré Flaugergues first detected it. The comet quickly brightened until it could be seen without a telescope, and it remained visible to the naked eye for 10 months. Its appearance was even credited with the coincidentally ultra-fine wines from that year. In October 1816 the English poet John Keats wrote of the thrill of reading a new work of literature:

Then felt I like some watcher of the skies
When a new planet swims into his ken;
Or like stout Cortez when with eagle eyes
He star'd at the Pacific...

Most critics assume that Keats was recalling Herschel's discovery of the planet Uranus in 1781, an event that occurred 14 years before he was born. I suspect the line harks back instead to the discovery of the magnificent comet of 1811, which appeared only 3 years before he wrote the sonnet. Other writers spoke of the sky too—here is Keats:

Bright Star: Were I as stedfast as thou art…

but Jonson's famous words about Shakespeare apply to Tennyson as well:

He was not of an age, but for all time!

In 1830, a young Alfred Tennyson published "*Poems: Chiefly Lyrical*," a selection that included his first published verse, "Supposed confessions of a Second-rate sensitive mind". It included this allusion to the night sky:

A grief not uninformed, and dull
Hearted with hope, of hope as full
As is the blood with life, or night
And a dark cloud with rich moonlight.

Tennyson reached the peak of his creative power in 1850 with the publication of *In Memoriam*, and his profound love of the night sky is the subject of the next chapter. This book then goes on to focus on Gerard Manley Hopkins, who made his first reference to a celestial body in his "Fragment of Anything You Like". The mood is light, and the poem offers a metaphor of an embarrassed moon away from her "traditional" place as "queen" of the night sky:

Fair, but of fairness as a vision dream'd:
Dry were her sad eyes that would fain have stream'd:
She stood before a light not hers, and seem'd

The lorn Moon, pale with piteous dismay,
Who rising late had miss'd her painful way
In wandering until broad light of day:

Then was discover'd in the pathless sky
White-faced, as one in sad assay to fly
Who asks not life but only place to die. (2C-P116)

Fourteen years later the subject of the Moon returns as the fragment "Moonrise". But this later fragment is entirely different. Instead of a metaphorical abstraction, the Moon is now the subject of an actual observation that Hopkins remembers:

I awoke in the Midsummer not to call night,
In the white and the walk of the mornings
The moon, dwindled and thinned to the fringe
of a fingernail held to the candle,
Or paring of paradaisical fruit, lovely in it waning but lustreless,
Stepped from the stool, drew back from the barrow,
of dark Maenefa the mountain;
A cusp still clasped him, a fluke yet fanged him,
entangled_him, not quit utterly.
This was the prized, the desirable sight, unsought, presented so easily,
Parted me leaf and leaf, divided me, eyelid and
Eyelid of slumber.

The “midsummer not to call night” in the first line recalls the night-long twilight that marks the summer solstice around the poem’s locale of St. Beuno’s: June 19, the date of writing, is only 2 days from the solistice. Hopkins writes that the time is early morning. Rising before the sun, the Moon is old and Hopkins records in the poem’s second line that it has “dwindled and thinned to the fringe of a fingernail”. The clasping and fanging by the cusp of the Moon suggests the metaphor of the Moon as an anchor embedded in the mountain; however the entanglement is not complete and leaves us wondering if “not quit utterly” should imply the moon sitting on top of Maenefa, flush with its slope, as if it is almost, but not quite, anchored. Possibly the Moon has already risen above Maenefa, but not above the distant trees growing on that mountain. Thus we could imagine the thin cusp of moon entangling in the trees.

This fragment shows a sense of accuracy in Hopkins’s sharp eyes that is seldom found in the passages by other authors. Hopkins was an astute observer of astronomical phenomena. The following chapters will test this idea, using both literary and scientific sources.

As an anecdote, one day in 1883 a young boy at Stonyhurst College was in a deserted playground nursing a sore tooth. The story he later passed on to Father Joseph Crehan relates how he was in some pain when “Father Hopkins appeared, walking from the Observatory (which is on the far side of the playground) towards the college. Father Hopkins came up to him and asked why he was there all alone. The boy explained, and then Hopkins said: ‘Watch me’. He took off his gown and proceeded to climb up one of the goalposts… Hopkins reached the top of the post and then lowered himself down. He put on his gown and then walked away. The boy’s excitement at this unusual activity made him forget his toothache, even though this was long before the age of psychiatrists.” That this curious incident took place as Hopkins was returning from a visit to Stonyhurst’s Observatory is significant for it indicates to us that Hopkins was not a mere casual observer of the heavens, but that he had sufficient interest (indeed we might conclude from the accuracy of the many references in his writings) to cause him to consult Father Steven Perry, the eminent astronomer who once directed the Stonyhurst Observatory. There are a number of matters that Hopkins could have been discussing during that visit to the Observatory. He had made detailed notes of some remarkable sunsets early that year and had written to the journal *Nature* about them, or he could have been discussing an astronomical observation he had made.

Although much research has been published on Hopkins’s concepts of God in nature and inscape, not much of it has concerned the quantity of astronomical knowledge, let alone the related subject of appreciation of the night sky, that Hopkins infused into his writings. Over one hundred images relating to astronomy brighten his works, and some of these have been carefully thought out so that both his special knowledge and ability to impart it add to their value. Thus, we can deduce from his references that he is an astute observer of astronomical phenomena. A comet with cocoon-like appearance approaching the sun, the Zodiacal light, several eclipses, and even some intriguing details of a moonrise at last crescent, are manifestations of this interest (Fig. 7.1).

Fig. 7.1 Moonrise. This image depicts a series of consecutive moonrise events observed during the Summer of 2007 from the site of the Adirondack Astronomy Retreat, not far from the town of Lewis, New York. Photograph by the author

Chapter 8
Arise and Fly

Come, gentle night; come, loving, black-brow'd night,
Give me my Romeo; and when he shall die,
Take him and cut him out in little stars,
And he will make the face of heaven so fine
That all the world will be in love with night,
And pay no worship to the garish sun.
Tho' much is taken, much abides; and tho'
We are not now that strength which in old days
Moved earth and heaven, that which we are, we are;
One equal temper of heroic hearts,
Made weak by time and fate, but strong in will
To strive, to seek, to find, and not to yield.

Tennyson acts as a bridge across the centuries between Shakespeare, the primary focus of my Ph.D. thesis, and Hopkins, the subject of my M.A. It took some effort for me to understand Tennyson's value as such a bridge. Accordingly, this chapter's opening passage combines two great pieces of poetry separated by 233 years. Both are in blank verse, both use the iambic pentameter foot, and both are phenomenal poetry. I did not credit the two passages on the same page deliberately, so that the reader can figure out where the first poem breaks and the second begins. As should be apparent, the part that ends with "garish sun" is from *Romeo and Juliet 3.2.20–25*. These lines are from the famous "balcony scene" although there was no balcony scene in the original Shakespeare; there was no such word in English at the time. David Garrick first staged it this way in the eighteenth century. The rest is from the closing six lines (ll. 65–70) of Tennyson's *Ulysses*. It is partly because of the similarity between the two poets, and poems, that I have chosen Tennyson to bridge the gap between Shakespeare and Hopkins. Since Tennyson was born before but died after Hopkins, the structure of his poetry relates well to that of Hopkins. Finally, Tennyson's own passion for astronomy helps to bring the fields together both in time and in content.

D.H. Levy, *The Starlight Night*, Astrophysics and Space Science Library 419,
DOI 10.1007/978-3-319-19878-1_8

Alfred Tennyson, the son of a rector in a town on England's East coast, was such an enthusiast that he owned a 2-inch diameter refractor telescope. Not only did he own one, but he also almost certainly used it frequently. According to Gillian Daw, Tennyson owned this refractor which he kept at his home in Aldworth in the Sussex region of England. Perhaps more important, he befriended some of the greatest English astronomers of his time, and he used their telescopes from time to time. Tennyson's most famous poem, *In Memoriam*, appeared in 1850, and a critic wrote that its most significant achievement was to "domesticate science" by expressing some of the new discoveries and ideas of the age in terms that were easy to understand. As a result, Tennyson would become one of nineteenth century science's greatest expositors. A man who captured the essence of the greatest scientific achievements of the nineteenth century, and then returned them to the world through the words and rhymes of his poetry, Tennyson became one of the great scientific teachers of his time.

During his years at Trinity College, Cambridge, Tennyson joined an intellectual group called the Apostles, and he thoroughly enjoyed the conversations he had with his friends there, particularly with his best friend Arthur Henry Hallam. Founded by then-student (later Bishop of Gibraltar) George Tomlinson, the Apostles was a secret society. By the start of the summer of 1833, Tennyson was feeling the pinch of some bad reviews about his first book, *Poems: Chiefly Lyrical*. The young poet was not adept at dealing with the negative criticism that included words like "obscure"; "he expresses his own peculiar character, by wishing himself to be something that he is not. … he wishes to be – a *river!* … Mr. Tennyson – himself, we presume, a dreamy lotus-eater, leaves them in full song." (Croker) Stung by this unfavorable bad review, Tennyson stopped publishing for a while. Then in August 1833, Tennyson received a letter from Hallam's uncle, Harry Eltron: "At the request of a most afflicted family," Elton wrote, "I write to you because they are unequal, from the grief into which they have fallen, to do it themselves." One can imagine the young poet holding his breath as he read on. "Your friend, Sir, and my much-loved nephew, Arthur Hallam, is no more." (Elton to Tennyson, September 1833). With the shattering news that Hallam had suffered a massive stroke, the young poet was shattered and stunned. He ascended the stairs of his home to share the news with his sister, who was also Hallam's fiancé. She fainted at hearing this news, and Tennyson fell into a deep depression and stopped writing for almost a decade.

One of Tennyson's initial thoughts after his friend's death was that his life was no longer worth living. The still-youthful Tennyson might have stopped writing at that time, but he did not stop studying. "The sudden extinction of his friend, with all his infinite capacity for affection and his brilliant promise, struck at the very roots of his will to live…. These months of suffering intensified the desire, which was to haunt the poet throughout his life, to find an answer to the great and insoluable questions, regarding the survival of the human spirit, the freedom of the human will, and the existence of a divine purpose guiding the universe." (Sir Charles Tennyson, 105). As the months following his friend's death turned into years, Tennyson still had no desire to publish his poetry but he did indulge his love of reading and studying. He used the time to good advantage by reading Charles Lyell's famous *Principles of Geology*, in which the broad outlines of the evolution of the Earth were first proposed—principles that would soon find a home in Tennyson's poems. He was also fascinated with the new theory about the descent of humanity from simpler species by Charles Darwin. Although Darwin did not publish his theory until 1850, the year of *In Memoriam*, Tennyson kept up with earlier publications that clearly were leading toward this theory. In addition, he immersed himself in the French astronomer Pierre-Simon Laplace's theory of the cosmogony of the solar system.

The Kant-Laplace Theory and the Solar Nebula Disk Model

A generation before Tennyson finished *In Memoriam*, Laplace published his outline of the Nebular Hypothesis in his treatise *Exposition du systeme du monde*. Struck by the similarity of motion and direction of the objects in our solar system, he proposed, in 1796, a theory that the Sun and planets condensed out of a cloud. As it rotated, it the cloud threw off a series of concentric rings, each of which condensed into a planet. Although it resembles an idea of stellar and planet formation that Immanuel Kant set forth in 1755 in *Universal Natural History and Theory of the Heavens*, the Laplace theory was developed independently and with far greater detail.

Coupled with Immanuel Kant's earlier suggestion, this model became known as the Kant-Laplace theory. With this work, Pierre-Simon Laplace was startlingly close to providing the understanding of the solar system's formation which persists today, despite issues with the Solar System's angular momentum that the nebular model could not explain. In Laplace's theory, developed from the philosophical writings by Immanuel Kant, the Sun formed out of a high-temperature cloud; as Tennyson would later describe the process as "ín tracts of fluent heat began." Now over 200 years old, the momentum distribution problem led Kant-Laplace theory to be supplanted in the early and mid-twentieth century by Thomas Chamberlin and Forest Moulton's planetesimal theory, James Jeans's tidal theory, which held that a passing star pulled material from the Sun to form the planets, and several other efforts including Otto Schmidt's accretion model, William McCrea's protoplanet theory, Michael Woolfson's capture theory, and Andrew Prentice's modern Laplacian theory. However for the past half century, as adapted in 1972 by Soviet astronomer Victor Safronov with the modern Solar Nebula Disk Model, it has been the accepted theory for the genesis of the solar system.

It does not take a great deal of effort to imagine such a large dark cloud in space, sitting passively for an incredibly long time. Known as a giant molecular cloud, it is more than 300 light years across—one of the biggest objects in the galaxy, yet almost invisible since there are no stars to light it. It is also one of the coldest, consisting of molecular hydrogen hovering near absolute zero. From a great distance, it might be visible as a dark nebula blocking our view of light from more distant suns. In the case of our cloud, other stars were situated closer to its edge; one of them, a massive star, filled with heavier atoms like carbon, suddenly ran out of material for nuclear fusion. In a fraction of a second the huge sun collapsed on itself, and then it blew itself apart in the extremely violent explosion of a supernova, spreading carbon into nearby space and into our giant molecular cloud. With the injection of carbon, and with the possible help of some ultraviolet irradiation, the cloud's particles heated up and began to recombine into organic substances.

Gradually a fragment of this cloud became unstable and started to collapse, the force of its own gravity causing it to coalesce into a thicker cloud, its tiny grains covered by a thin layer of organic molecules like frosting on a cake. Possibly the shock wave from the supernova itself set it off, or for some other reason the cloud began to compress. It also started to rotate slowly, and gradually the grains settled in the nebula's central plane. The particles clumped together, or accreted, very slowly as individual particles that settled to the central plane found one another. At the center of the accretion disk was a body that would grow and grow, all the while continuing to get hotter.

During this ancient time, smaller clumps broke off the central cloud. As the clumps ran into each other, they grew into larger bodies called planetesimals, and into smaller bodies some call protocomets or cometesimals. Alternatively, the planetesimals grew bigger until many collisions from smaller objects started to fragment the primordial bodies. As temperatures soared, the delicate organic matter was at risk. Unable to survive the intense heat, it began to disintegrate. Meanwhile the outer part of the accretion disk, which did not get as hot, retained some of its organic materials. Finally, the temperature at the cloud's center grew hot enough to ignite, and our Sun was born as a star and began nuclear fusion.

The entire process between the collapse of the molecular cloud, the building of the planets, and the ignition of the Sun should not have lasted more than 100 million years. The big gas giants Jupiter and Saturn had to have accreted before the Sun turned on its fusion furnace, for soon after that the remaining gas in the primordial nebula would have been blown away. Uranus and Neptune were probably the last to grow. Once formed, the infant solar system had a group of warm inner planets without a large amount of organic material, and some cold outer planets where organic materials remained. These particles, specifically carbon, hydrogen, oxygen, and nitrogen, form the modest but eloquent alphabet of life, and as part of distant comets and asteroids, they found their way back into the inner solar system. Most likely they were deposited on Earth through the collisions by comets and asteroids.

What I have just explained is the updated version of Kant-Laplace, now commonly known as the Solar Nebula Disk Model. On a much smaller scale, solar-system sized dark clouds could evolve out of a giant cloud. These clouds are now known as Bok globules, after the astronomer who first described them. Bart Bok and his associate Edith Reilly found some 200 of them during a preliminary survey and published their theory in 1947. The globules were typically round, about one-sixth the apparent diameter of the Moon in the sky. They are extremely dense, so thick that if one of them were to cover the zero-magnitude star Vega, even the Hubble Space Telescope, which can penetrate down almost to thirtieth magnitude, would not be able to spot the hidden star. "Through a telescope you could come to the leading edge of one of these things and suddenly the stars would just disappear," Bok once told me. "And then you would push the telescope's slow motion button a bit and bloop! The stars come back." (Levy, 64) These clouds, Bok believed, marked the birthplaces of new stars. After staying together in space for untold millions of years, they begin to collapse under their own gravity. As they get denser the process of contraction continues until the center ignites, stellar fusion begins, and a new star is born. Bok and Reilly had stumbled onto a gem of stellar physics.

As Bok tried to locate more of these objects and study them further, he also set about looking for a suitable name. The answer came to him one morning as he opened the front door of his home to pick up the freshly delivered glass bottles of milk. In those days during the 1950s, milk was not homogenized, so the top of each bottle sported a small layer of cream. As Bok set one of the bottles down in the kitchen, he studied the tiny globules of fat floating around the cream layer. "These look just like my globules!" he suddenly realized.

In Tennyson's age, however, the question of the nature of deep sky objects (a more generic term at the time) was in its infancy. The objects described by Charles Messier, the famous comet discoverer who cataloged more than 100 celestial objects as part of his search of the night sky, and many others listed by William Herschel, were often interpreted as youthful planetary systems. It was possible, some thought, that the nebulae that were being discovered across the sky were distant examples of solar systems in formation. Only in recent history have the supposed nebulae of the nineteenth century been unveiled for what they really are: clusters of stars, clouds of gas and dust, and the distant galaxies of our expanding universe.

Turning Science into Poetry

After long periods of reading and study, Tennyson attempted to place the original Kant-Laplace Solar System formation theory, explained in detail above, into his poetic consciousness. Tennyson immersed himself so deeply into these ideas that one might have suspected that he was thinking of becoming a scientist in lieu of engaging in poetry, and the scientists he did befriend were probably impressed with his ability to grasp the nebular hypothesis so effectively. Formulated by Pierre Simon, Marquis de Laplace, this hypothesis held that the planets are children of the Sun, formed of the outskirts of the cloud whose center became the Sun. If that theory is correct, then the Sun is slowly losing its heat and will someday die. While that will happen eventually, the Sun will retain its current basic state for yet another 5 billion years.

By the 1840s, Tennyson let his lively interest in the night sky permeate his writing, as we see in this song from his 1847 poem *The Princess*:

> Now lies the Earth all Danae to the stars,
> And all thy heart lies open unto me.
> Now slides the silent meteor on, and leaves
> A shining furrow, as thy thoughts in me.

During these years, quatrain by quatrain, thought by thought, Tennyson was slowly building his masterpiece. It evolved into something far greater than a remembrance of his friend. It was a celebration of mid-nineteenth century life, a work strengthened by its author's own passion for the natural world.

Tennyson used the facts of science to enrich the language of poetry, as in the following example:

Who loves not Knowledge? Who shall rail
Against her beauty? May she mix
With men and prosper! Who shall fix
Her pillars? Let her work prevail. (CXIV, ll.1–4)

Who indeed should set the pillars of Hercules, the limits of knowledge? Six cantos later, Tennyson suggests an answer:

Let Science prove we are, and then,
What matters Science unto men,
At least to me. I would not stay. (CXX, ll. 6–8).

To understand the powerful role of science, particularly astronomy and evolution, in this poem, we need to know that "it is the story of a soul stunned by a tremendous loss," notes one critic, "and struggling to find the meaning of a universe in which such losses can occur" (Victorian Poetry and Poetics, 78.) How common a sentiment. We have all have suffered tragic losses, and when we do we tend to question the natural order that allows such things to happen. *In Memoriam* goes further by reflecting the strength of its author's love of careful research into the sciences and theology. "He lacked and distrusted the passion of the revolutionary," this critic notes, "for creating and imposing exclusive ideas of truth by skipping inconvenient evidence." (Pearce, xl.) Tennyson insisted on submitting the ideas of the emerging scientific thought "to the test of his own feelings and his own vision."

At the start of *In Memoriam*, Tennyson puts into words his grief at the loss of his friend, but at the same time he wished to write about a Universe that brings such grief. The words from his pen are an imperfect tool for this purpose, but Nature seems to him to be structured partly to give answers, but mostly to raise questions.

I sometimes hold it half a sin
To put in words the grief I feel;
For words, like Nature, half reveal
And half conceal the Soul within. (V, ll.1–4)

Understanding that wordcraft is the poet's basic tool, Tennyson's pen is strong enough to suggest a vision of the Universe. This vision sees the Earth both as the familiar world we live in, and as a small body in space.

This round of green, this orb of flame,
Fantastic beauty; such as lurks
In some wild poet, when he works
Without a conscience or an aim. (XXXIV, ll. 5–8)

As the epic poem twists and turns, its words occasionally paint a picture of a natural scene that takes this "fantastic beauty" and puts it to work:

We talk'd: the stream beneath us ran,
The wine flask lying couch'd in moss,

Or cool'd within the glooming wave;
And last, returning from afar,
Before the crimson-circled star
Had fallen into her father's grave,

And brushing ankle-deep in flowers,
We heard behind the woodbine veil
The milk that bubbled in the pail,
And buzzings of the honeyed hours. (LXXXIX, ll. 43–52)

The Sun has set at the end of the day, and with the end of twilight a bright planet (the crimson_circled star) is following her father, the Sun, in setting. This planet is almost certainly Venus, as an evening star.

Yet *In Memoriam* is far more than a description of Nature's elegant natural beauty. It is a serious attempt to look at the scientific world of Tennyson's time, and it deals with scientific issues in remarkable detail. In the following passage, Sorrow, the priestess, suggests a pattern for the Universe:

O Sorrow, cruel fellowship,
O Priestess in the vaults of Death,
O sweet and bitter in a breath,
What whispers from thy lying lip?

"The stars," she whispers, "blindly run;
A web is woven across the sky;
From out waste places comes a cry,
And murmurs from the dying sun;

And all the phantom, Nature, stands--
With all the music in her tone,
A hollow echo of my own--
A hollow form with empty hands."

And shall I take a thing so blind,
Embrace her as my natural good;
Or crush her, like a vice of blood,
Upon the threshold of the mind? (III, ll. 1–16)

Tennyson mistrusts a Universe which seems to follow a random course with no order or pattern. It is a similar mistrust to what John Davies noted in *Orchestra* (see Chap. 2), by placing the new cosmology in parentheses so that it would not interfere with the carefully ordered cosmic dance of Orchestra. The new cosmology seems a random, cold, and silent universe in which humanity's prayers are unheard:

Man, her last work, who seemed so fair,
Such splendid purpose in his eyes,
Who roll'd the psalm to wintry skies,
Who built him fanes of fruitless prayer,

Who trusted God was love indeed
And love Creation's final law--
Tho' Nature, red in truth and claw
With ravine, shriek'd against his creed--

Who lov'd, who suffer'd countless ills
Who battled for the True, the Just,
Be blown about the desert dust,
Or seal'd within the iron hills? (LVI, ll. 9–20)

Sixteen years after the publication of *In Memoriam*, Gerard Manley Hopkins, a poet whose work we will explore in the next chapters, probed in *Nondum* the mysteries of the night sky and asked:

> God, though to Thee our psalm we raise
> No answering voice comes from the skies;
> To Thee the trembling sinner prays
> But no forgiving voice replies;
> Our prayer seems lost in desert ways,
> Our hymn in the vast silence dies. (Gardner and MacKenzie, *The Poems of Gerard Manley Hopkins, 4th ed.,* 32.)

Both poets wrote from a strong spiritual sense, strengthened with a profound wonder at the sky. The cry from a dying Sun invoked by Tennyson in the stanza beginning, "The stars … blindly run" harks back to the nebular hypothesis. What I find more interesting than this idea is the one expressed in the first statement from "Sorrow" that the stars blindly run to weave a web across the sky. In one interpretation of "blindly," Tennyson suggests that objects in the sky do not obey an intelligence but the celestial mechanics worked out by Newton, whose laws of gravitation described how objects travel through space. Yet this interpretation does not seem fully plausible. "Sorrow" describes a celestial system where nothing works right, and the blind running of the stars is meant to suggest cosmic anarchy.

References to the nebular hypothesis appear throughout *In Memoriam*, even at places where it is invoked peripherally, as with the eddies of gas involved in the formation of the planets, appearing again in Canto 128:

> No doubt vast eddies in the flood
> Of onward time shall yet be made (CXXVIII, ll. 5–6)

Laplace's theory suggested an ordered Universe, but Tennyson's "Sorrow" still saw stars blindly running, operating without logic. In the years before the completion of *In Memoriam*, several bright comets had "blindly run" their courses, in a sense, to weave a web across the sky. The Great Comet of 1811, which will be discussed later, was one such comet. Eight years later, a second bright comet rounded the Sun. However, the appearance of a great comet in 1830, just 3 years before Tennyson first penned this group of stanzas, and another in 1831—two great comets in as many years—was highly unusual and might have added credence to Tennyson's image of objects blindly crossing the sky. Although the first of these two comets attracted some attention, when it was brightest the 1830 comet was visible only from the southern hemisphere, heading north. The second comet appeared a few months later, and was first seen over England with a bright, short tail. Halley's comet passed over the sky of Earth in 1835, and a magnificent comet appeared in 1843. It ruled the evening sky throughout the world that March, and as it completed its hairpin encounter with the Sun, it grew a tail that stretched for 70 million miles, almost as long as the distance between the Sun and Earth.

In 1846, a fainter comet appeared, a periodic comet whose course astronomers thought they understood well until as it moved through the sky it split into two pieces. At the time, a comet splitting into two was an unknown event, difficult to explain, one which could have inspired a thought about objects blindly running about the sky. Comets are now known to split apart with some frequency, and these events are sometimes dramatic. By far the most spectacular example of a comet's breaking up occurred in July, 1992, when, out of Earth's sight, Comet Shoemaker-Levy 9 broke apart as a result of a close encounter with Jupiter. When the comet was discovered in March, 1993, its 21 pieces were described as a string of pearls. The comet fragments collided with Jupiter in July, 1994.

Other major events were also taking place in astronomy at the time. In 1846, two astronomers, John Couch Adams from England, and Urbain Leverrier from France, independently calculated where the eighth planet should lie. Finally, Johann Galle and Heinrich d'Arrest in Germany turned their telescope to the proper location and discovered the planet, Neptune. Just 4 years later, Tennyson alluded to the event, with "arms" referring to telescopes:

A time to sicken and to swoon,
When Science reaches forth her arms
To feel from world to world, and charms
Her secret from the latest moon? (XXI, ll. 17–20)

Besides Neptune, which was hardly a moon, some notable new and real moons were discovered during the middle years of the nineteenth century. William Lassell discovered Triton, orbiting Neptune, in 1846, the year of Neptune's discovery. Hyperion, the seventh of Saturn's moons, was detected by Harvard's George Bond in 1848, 2 years before the publication of *In Memoriam*. Ariel and Umbriel were found orbiting Uranus in 1851, by William Lassell. Also, between 1801 and 1849 the first ten asteroids were detected. This was the new science, a litany of observations and ideas that were changing the way we look at ourselves, like the sunspots which wander like isles of night across the blinding Sun. In the next lines Tennyson explains, in the most profoundly moving way, the majesty of sunspot activity:

And was the day of my delight
As pure and perfect as I say?
The very source and fount of day
Is dash'd with wandering isles of night. (XXI, 17–20)

The process of scientific discovery is a slow and delicate one. Although Galileo first detected sunspots in 1610, observers watched the Sun for another 200 years before Heinrich Schwabe, an amateur astronomer from Dessau, Germany, bought a small refractor telescope in 1826. With it he began to observe the Sun in hopes of finding the elusive planet Vulcan crossing in front of the solar surface. In a beautiful illustration of serendipity in astronomy, Schwabe never found Vulcan (the planet-sized world does not exist), but in 1843, he announced the existence of a cycle of sunspot activity lasting about a decade.

The discovery of Neptune was evolving at the same time. At first the calculations of neither Adams nor Leverrier were taken seriously. George Airy, England's astronomer royal, simply did not believe Adams, and LeVerrier had to find astronomers on his own—Johanne Galle and Heinrich d'Arrest—who were willing to conduct the search that led to the discovery. It is the beautiful process of discovery and learning that Tennyson refers to in Canto 23:

When each by turns was guide to each,
And Fancy light from Fancy caught,
And Thought leapt out to wed with Thought
Ere Thought could wed itself with Speech (XXIV, ll. 1–4)

Wherever humanity stands, our scientific understanding is just at a beginning. In one of *In Memoriam's* most exquisite passages, Tennyson looks upon humankind as babes wandering about in the woods, insisting that we do not yet know how to think of ourselves in the Universe:

O, yet we trust that somehow good
Will be the final goal of ill,
To pangs of nature, sins of will,
Defects of doubt, and taints of blood;

That nothing walks with aimless feet;
That not one life shall be destroy'd,
Or cast as rubbish to the void,
When God hath made the pile complete;

That not a worm is cloven in vain,
That not a moth with vain desire
Is shrivell'd in a fruitless fire,
Or but subserves another's gain.

Behold, we know not anything;
I can but trust that good shall fall
At last far off at last, to all,
And every winter change to spring.

So runs my dream, but what am I?
An infant crying in the night;
An infant crying for the light,
And with no language but a cry (LIV, ll. 1–20)

This powerful thought reflects the emotions of some of us who have spent years gazing at the sky, or studying some other aspect of nature. On a clear night the sky can seem peaceful, even benevolent. It is hard to avoid connecting a starry night with the thought that all is right with the universe, and that good shall ultimately triumph, and that every winter shall indeed turn to spring. But then comes the reality check—Nature is neutral, and we cannot begin to understand the meaning, if there is one, of a dark and lovely night. Humans are literally still infants crying in this night, in a Universe so complex and vast that we can't even put the question into proper words. A more modern expression of this idea comes from the film *Star Trek: The Motion Picture* in which Mr. Spock, played by Leonard Nimoy, uttered "It knows … only that it needs, but like most of us, it does not know what."

Canto 118 gives poetic voice to the biological breakthrough that was taking place at the very same period that Tennyson was composing *In Memoriam*. A decade after the publication of Tennysonn's epic, Charles Darwin's studies on the Galapagos Islands culminated in 1959s *The Origin of Species*, his great contribution to our understanding of our heritage and ancestry. Darwin's work (accessible to those in his circle even before the publication of his opus) led to the idea that life forms evolve from earlier forms, and that the Earth itself has evolved though time:

Contemplate all this work of Time,
The giant laboring in his youth;
Nor dream of human love and truth,
As dying Nature's earth and lime;

But trust that those we call the dead
Are breathers of an ampler day
For ever nobler ends. They say,
The solid earth whereon we tread

In tracts of fluent heat began,
And grew to seeming_random forms,
The seeming prey of cyclic storms,
Till at the last arose the man;

Who throve and branch'd from clime to clime,
The herald of a higher race,
And of himself in higher place,
If so he type this work of time

Within himself, from one to more;
Or, crown'd with attributes of woe
Like glories, move his course, and show
That life is not as idle ore,

But iron dug from central gloom,
And heated hot with burning fears,
And dipt in baths of hissing tears,
And batter'd with the shocks of doom

To shape and use. Arise and fly
The reeling Faun, the sensual feast;

Move upward, working out the beast,
And let the ape and tiger die. (CXVIII, ll. 1–28)

Tennyson's enthusiasm about the idea of evolution is tempered by his belief that some teleological force has control over the process, letting the lower life forms fall away. Noting that the ape and tiger die, Tennyson crowns evolution with a sense of divine purpose; the evolution of the Earth is reproduced by the development of life. This is a theme that recurs in his other poetry, for example in Maud, which appeared 5 years after *In Memoriam*:

So many a million of ages have gone to the making of man:
He now is first, but is he the last? is he not too base?

The man of science himself is fonder of glory, and vain,
An eye well_practised in nature, a spirit bounded and poor;
The passionate heart of the poet is whirl'd into folly and vice.

I would not marvel at either, but keep a temperate brain;
For not to desire or admire, if a man could learn it, were more
Than to walk all day like the sultan of old in a garden of spice. (Maud, lines 136–143)

In staking out the grounds of his theory of evolution, Charles Darwin proposed two ideas. One was evolution itself, which says that humans moved upward, working out the beast. The second was natural selection, Darwin's proposed mechanism to explain how evolution takes place. While scientists generally agree on the first, there remains hot debate about the second. According to Darwin's basic theory of natural selection, species evolve gradually, without sudden change. However, evidence now exists for a series of bursts of new species. A comet or asteroid impact, or a series of damaging volcanic eruptions, could trigger such a burst of speciation.

The best known such impact took place 65 million years ago. In just a few minutes, an Earth teeming with many forms of life was turned into a burning wasteland. The change that occurred at the end of the Cretaceous period of Earth's history was anything but gradual. With large thunderclaps and a huge crash, an asteroid (or possibly a comet) slammed into the Earth at the eastern edge of what is now Mexico's Yuçatan Peninsula. Walls of water a mile high or more raced out from the point of impact, and millions of tons of dust surged upwards in a gigantic cloud. A crater at least 100 kilometers wide and several miles deep was formed in less than a minute, and excavated material from the crater rushed out with such force that it quickly circled the Earth. As the Earth's surface was blitzed with debris, temperatures rose as high as those in a broiling oven, setting off a worldwide firestorm. Within a few weeks, the whole planet was shrouded in a cloud of dust and soot. The sky was absolutely black, and for over a month there was no sunlight whatsoever, anywhere on Earth. The rain was dense with sulfuric acid. Finally, as the clouds dissipated, Earth was left with a global warming period that lasted for centuries.

By the time the episode was over, more than three quarters of the species of life on Earth had vanished. Although Tennyson was not aware of Earth's great cataclysm, he did know that at one time dinosaurs roamed the Earth, and that they were now extinct:

There rolls the deep where grew the tree.
O earth, what changes hast thou seen!
There where the long street roars hath been
The stillness of the central sea.

The hills are shadows, and they flow
From form to form, and nothing stands;
They melt like mist, the solid lands,
Like clouds they shape themselves and go. (CXXIII, 1–8).

The only thing unchanging about the Earth is that it is always changing. Where once there were seas, mountains appear, and the mountains of long ago eroded down to hills and valleys. Our understanding of this continuity is not new; even the author of *Ecclesiastes* wrote eloquently of the majesty of Earth's water cycle, and of other cycles of change in Earth's history:

All the rivers run into the sea; yet the sea is not full; unto the place from whence the rivers come, thither they return again.

The thing that hath been, it is that which shall be; and that which is done is that which shall be done: and there is no new thing under the sun. (Ecclesiastes 1:7.)

Although some sources suggest that Solomon himself authored *Ecclesiastes*, Martin Luther and others date the book to a later author. Possibly the book had a Solomonic origin, with the famous king writing a portion of it and a later writer completing it, but since fragments of the text were found at Qumran in the Dead Sea Scrolls, it could not have been dated later than 150 BCE. (See C. Ryrie, *The Ryrie Study Bible* (Chicago: Moody Press, 1976.))

In Canto 121, Tennyson enjoys a poetic invocation of Hesper and Phosphor, the ancient Greek names for Venus as the evening star and the morning star. Bringing epic tradition up to date means asserting that the two stars are really a single planet, Venus. As Hesper, Venus never sets long after the Sun. Tennyson knows that Phosphor can never rise during the same night; Venus is either one or the other for periods of several months at a time. But when Phosphor does rise always in the pre-dawn sky, the "greater light" of the Sun cannot be far behind.

> Sad Hesper o'er the buried sun
> And ready, thou, to die with him,
> Thou watchest all things ever dim
> And dimmer, and a glory done.
>
> …
>
> Bright Phosphor, fresher for the night,
> By thee the world's great work is heard
> Beginning, and the wakeful bird;
> Behind thee comes the greater light.

Finally, Tennyson reveals, in the climax of this canto that the dual evening and morning star are really one:

> Sweet Hesper_Phosphor, double name
> For what is one, the first, the last,
> Thou, like my present and my past,
> Thy place is changed; thou art the same. (CXXI, ll. 1–4, 9–12, 17–20.)

Again and again one can see that Alfred, Lord Tennyson was a poet whose works were strengthened by repeated references to the natural world. The beauty of the night sky, as his 1842 poem *Locksley Hall* so effectively shows, was a central part of that marriage between poetry and the night sky:

> Many a night from yonder ivied casement, ere I went to rest,
> Did I look on great Orion sloping slowly to the west.
>
> Many a night I saw the Pleiads, rising thro' the mellow shade,
> Glitter like a swarm of fire_flies tangled in a silver braid. (ll. 7–10).

As much as it was one of Tennyson's greatest poetic achievements, *In Memoriam* was a scientific achievement as well, for it presented the accomplishments of the science of the time to a wide readership of culturally sophisticated people who were not scientists. But to claim that *In Memoriam* is a mere poetic explanation of science is to limit its overwhelming power. The opus did not seek to explain the scientific accomplishments of the day so much as to humanize and romanticize the scientific accomplishments of the Victorian age. Although many poets wrote of the night sky, it was Tennyson who tried to connect the romance of the sky with the accompanying scientific theories, and in doing so, he created a poetic legacy that helped to change the focus of later writing.

Is the sky a means by which we humans can see the extent of our dreams, the limit of what is possible, and even the limit of the horizon of our soul, a limit that knows no bounds? Is this the major discovery of those who share a passion for the night sky? Perhaps. We give the second-to-last word to Matthew Arnold, who suggests that the function of poetry is to ring out this thought:

> Plainness and clearness without shadow of stain!
> Clearness divine!
> Ye heavens, whose pure dark regions have no sign
> Of languor, though so calm, and, though so great,
> Are yet untroubled and yet unpassionate;
> Who, though so noble, share in the world's toil,
> And, though so task'd, keep free from dust and soil!
> I will not say that your mild deeps retain
> A tinge, it may be, of their silent pain
> Who have long'd deeply once, and long'd in vain __
> But I will rather say that you remain
> A world above man's head, to let him see
> How boundless might his soul's horizons be,
> How vast, yet of what clear transparency!
> How it were good to abide there, and breathe free;
> How fair a lot to fill
> Is left to each man still! (Arnold, "A Summer Night" lines 76–92.)

Influence from Friends

Tennyson's wide interests were nurtured and encouraged by his many friends, not the least of which was Edward FitzGerald, whose own interest in astronomy manifested itself through the twelfth century Persian astronomer whose *Rubáiyát* was a call to action for all those who loved the sky. The original twelfth century poem emphasized the joy of drink as much as it did the elation awarded by the night sky:

> Wake! For the Sun, who scattered into flight
> The Stars before him from the Field of Night,
> Drives Night along with them from Heav'n,
> The Sultan's Turret with a Shaft of Light.
>
> Before the phantom of False morning died,
> Methought a Voice within the Tavern cried,
> "When all the Temple is prepared within,
> Why nods the drowsy worshipper outside?" (ll. 1–8.)

In our modern age, how many of us have pleaded guilty to the crime of ignoring a beautiful night? Possibly one can be forgiven if there is an appreciation of what follows in this wonderful poetry, in my mind a description of an eclipse of the Sun. FitzGerald begins with a passing reference to a shadow:

> And Hell the shadow from a Soul on Fire

And a few lines down:

> We are no other than a moving row
> Of Magic Shadow-shapes that come and go
> Round with the Sun-illumnined Lantern held
> In Midnight by the Master of the Show.

FitzGerald is now prepared to complete the metaphor:

> The great shadow is from the Moon as our neighbor world crosses the Sun like a finger.

And in stanza LXXI the metaphor is completed:

> The Moving Finger writes; and, having writ,
> Moves on: nor all your Piety nor Wit
> Shall lure it back to cancel half a Line,
> Nor all your Tears wash out a Word of it.

Supporting evidence for this interpretation comes from the autobiography of the amateur astronomer Leslie C. Peltier, an unlikely source indeed. He brings the eclipse and moon shadow image literally to the printed page:

> All over America the eclipse was ended. "The moving finger writes and, having writ, moves on." Like a moving finger of darkness then cone-shaped shadow of the moon had dipped down, scrawled its brief two-minute mark of night across the land and then moved on, still writing, but now with invisible ink upon the empty page of space. (Peltier, 94.)

Old Fitz, as his friends, including Tennyson, called him, loved to write, but aside from the *Rubáiyát* the vast majority of his writings are preserved only as letters.

This poem was well known to Tennyson, who wrote "old Fitz" several times to express his joy at reading it:

> Than which I know no version done
> In English more divinely well;
> A planet equal to the sun
> Which cast it. (Tennyson, 1911)

Their friendship lasted for nearly 50 years. It was close and constant, but consisted mostly of correspondence between the two poets, one who wrote but a single successful poem, the other with a lifetime of composition. But unlike many authors, Tennyson was quite forgiving in his criticism.

In 1850, a year that could clearly be entitled Tennyson's *annus mirabilis*, *In Memoriam* appeared and was popular instantly. In that same year he was married and was named poet laureate. Tennyson's frequent allusions to Darwin's theories of evolution and natural selection did not raise the controversy then that they surely would today. *In Memoriam* began innocuously as a series of quatrains written in memory of Tennyson's friend, but it turned out to be far more than that, vastly greater than an elegy to Hallam. One of the truly great poems of English Literature,

In Memoriam begins as a memorial but ends, through a poetic description of his sister's Cecilia's wedding to Edmund Lushington, as a celebration of mid-nineteenth century life, and of the unprecedented scientific advances of that time. As critics Houghton and Stange suggested, the poem's most profound achievement was to domesticate science, to bring it to the level of a wide readership, and to attach to it a moral and religious interpretation. It represents the high-water mark for Tennyson's enthusiasm for introducing scientific concepts into poetry, both in the amount of scientific information presented and in its poetic effectiveness.

The Closing of In Memoriam

In Memoriam closes on a pantheistic note, specifically in the last line of the penultimate stanza. In a deeply personal phrase Tennyson returns one last time to Hallam:

> That friend of mine who lives in God,

From this warm and personal final mention of Hallam, he asks and answers the question What God? In what kind of home does Hallam reside? He begins his answer in the very next line, where Hallam's home, God, is reintroduced as "ever" immortal and loving:

> That God, which ever lives and loves,

is a pantheistic deity which lives everywhere and whose Universe is destined to some ultimate goal. The line is a teleological conception of the poet's belief in such an ordered Universe. The idea expands significantly in the next six words:

> One God, one law, one element,

completes the transformative poetical view from humanity to the complex interplay between science and religion. "One God" appears to represent Tennyson's understanding of the Judeo-Christian definition of a unique deity, and to it he adds the teleology that nature's laws proceed to some end result. In so stating, Tennyson anticipates an idea held by most modern physicists, that the universe and its particles can be explained by a single, "unified" theory. A solitary element follows from the one law, not one chemical element but a simple way of understanding the Universe. "The most incomprehensible thing about the Universe," Einstein said three quarters of a century later, "is that it is comprehensible." In our time, Stephen Hawking's lifetime effort to formulate a unified "theory of everything" brings back to physics what Tennyson brought to poetry.

> And one far-off divine event,

soars poetically to the ultimate end of the Universe. This makes some sense; if nature's small acts have specific results, then the ultimate, cumulative act that is the whole of nature must point to the end of the Universe, or at least its close as we understand it. Cosmologists ponder these possibilities: if it has sufficient mass, then the Universe will slow its expansion until it stops, and then a contraction will begin and continue until all the material in the Universe is compressed into a single point, as it was some 13.7 billion years ago. If there is not enough mass to stop the expansion, there will be a day when the last supernova explodes, and when, eons later, the Universe can no longer support even the simplest atomic matter. It is indeed one of these two faroff events

To which the whole creation moves.

This final line is an invitation to see the Universe, but not with a telescope. To visualize what Tennyson has in mind, we are asked to put the telescope aside and stare. It is only with an all-sky look with the unaided eye, particularly when the sky is clear and dark, far from light pollution. His answer, and ours, lies in the darkness of the night sky. Do not study it, do not learn it, just appreciate it, just gaze at it. Tennyson wrote it in a single, crashing line, and another poet, Ralph Hodgson, would echo the same thought in several lines more than 60 years later:

I stood and stared, the sky was lit,

The sky was stars all over it.

I stood, I knew not why,

Without a wish, without a will,

I stood upon that silent hill,

And stared into the sky, until

My eyes were blind with stars, and still

I stared into the sky. (Ralph Hodgson, "The Song of honour", closing lines.)

Tennyson's closing stanza of his greatest work of art seems a mystery until the reader takes this broader look not at a detail of the sky, like a planetary world, but at the entire sky. Just as one can read a poem, one can read any work of art, whether it be a painting, a work of sculpture, a great mountain range, or the night sky. By looking at the sky without a telescope, one can read the entire vault of night. Without a telescope or binoculars, on a clear dark night it is possible to see as many as three thousand stars at once. These stars do remind us of Tennyson's deeply personal elegy to his friend, as eulogized in the work's final verse:

That friend of mine who lives in God,

That God, who ever lives and loves,

One God, one Law, one element,

And one far-off divine event,

To which the whole creation moves. (CXXXI, ll.140–144)

Tennyson and His Telescope

Tennyson's affiliation for the night sky, if not a love, surely was at least an affection. Sir Norman Lockyer wrote that Tennyson enjoyed viewing through the 6-inch diameter Cooke refractor. "He was therefore often in the observatory." (H. Tennyson, 286.) At his Aldworth home he had his own 2-inch telescope with which he enjoyed views of the Moon and planets, double stars, and occasionally star clusters and nebulae. His son wrote that "there was a 2-inch telescope at Aldworth… The last time I met him he would talk of nothing but the possible ages of sun and earth, and was eager to know to which estimates scientific opinion was then veering." (Hallam Tennyson, 1911) He observed most of the comets of his time, including the Great Comet of 1854. "My father was always interested in the imaginative views which his children took of our surroundings. Of these I may give one instance: how Lionel had been brought from his bed at night, wrapt in a blanket, to see the great comet, and suddenly awaking and looking out at the starry night, asked, "Am I dead?"" (Vol. I, 370). He also witnessed Donati's Comet in 1858, specifically noting the great comet's central coma passing near Arcturus. (H. Tennyson, 308). He casually noted the passage of the Great Comet of 1861: "On our arrival at Clermont" Hallam wrote, "the comet was flaring over the market-place." (H. Tennyson, Vol. I, 472–473) and finally he alludes possibly to Coggia's great comet of 1877. (H. Tennyson, Vol. 2, 186) As a teenager he understood the finite speed of light, and how "The rays of many a rolling central star / Aye flashing earthwards, have not reached us yet." (Vol. I, 20)

Tennyson has been gone for more than a century, yet his poetry lives on, probably more popular now than it was during his lifetime. We do know that at least some of the Pre-Rapheaelites had qualms with Tennyson: "I find he has (*valeat quantum*) an unfavourable impression as to the character of Tennyson, even as a poet: he regards him as selfish, narrow in money-matters, not of lively affections: he is punctilious in pasying his score in company, and expecting his companions to pay theirs." (Rossetti, 333.) Such criticisms were even extended by his friend Browning, who allegedly "deprecates the publication by Tennyson of the trifling affairs… he says that T's books are declining in sale within this year or two (perhaps the influence of Swinburne.)" (Rossetti, 302.)

Tennyson shared an interest in the medieval period with William Morris, another prominent Pre-Rapheaelite. Although there are not many night sky interpretations in his writing, there is a moving poem called *Summer Dawn*.

> Pray but one prayer for me 'twixt thy closed lips,
>
> Think but one thought of me up in the stars.
>
> The summer night waneth, the morning light slips,
>
> Faint and grey 'twixt the leaves of the aspen,
>
> Betwixt the cloud-bars,
>
> That are patiently waiting there for the dawn
>
> Patient and colourless, though Heaven's gold

Waits to float through them along with the sun. ("Summer Dawn")

For someone who did not apparently display any interest in the sky, Morris describes this particular summer dawn perfectly. Night does not vanish, but it "waneth" gradually as thin bars of clouds cross the sky, "patient and colourless" and lazily awaiting the onset of day.

Tennyson Today

Even though Tennyson's poetry might seem dated now, it has withstood the time very well. When I first studied *In Memoriam* in 1969, I was inspired by his knowledge of science matters, and that surprise has grown to astonishment over the years. Tennyson's curiosity in the natural world should not be surprising, but his specific focus on the night sky is. It even appears to have been passed down to Tennyson's living great-great grandson, Jonathan Tennyson, an astrophysicist based in England. He has developed an interesting career in his own right, having published many papers, two of which deal with Comet Shoemaker-Levy 9's impact with Jupiter in 1994. Tennyson wrote to me in 2014 about what inspired him to go into science. I asked if his great-great-grandfather's poetry might have triggered such an interest. "I have to say not really or possibly the reverse. My great-great grandfather cast (casts?) a rather long shadow down the generations. Somehow it is more comfortable being a scientist." (J. Tennyson to Levy, 3 February 2014.) So it was that on the morning of November 12, 2014, I visited Dr. Tennyson at the University College London. The family connection was instantly obvious the moment I entered his office, for on its wall hung a framed excerpt from *In* Memoriam, his ancestor's greatest gift to the world:

There rolls the deep, where grew the tree.
O Earth, what changes hast thou seen!
There, where the long street roars, hath been,
The stillness of the central sea.

The hills are shadows, and they flow
From form to form, and nothing stands;
They melt like mists, the solid lands,
Like clouds they shape themselves and go. (*In Memoriam* CXXIII, 1–8)

Not long before his death in 1892, Tennyson asked his son Hallam that all future collections of his work end with the one poem that acts as his own obituary, "Crossing the Bar." It is appropriate also that there is a reference to the sky in the first line, most likely to glorious Venus which he loved to see and of which he wrote so often. Since this particular chapter offers a selection of his poetry that related most closely to the wonder of the night sky, I close this particular chapter by honoring his request:

Sunset and evening star,
And one clear call for me!

And may there be no moaning of the bar,
When I put out to sea.

But such a tide as moving seems asleep,
To full for sound and foam,
When that which drew from out the boundless deep
Turns again home.

Twilight and evening bell,
And after that the dark!
And may there be no sadness of farewell,
When I embark;

For though from out our bourne of Time and Place
The flood may bear me far,
I hope to see my pilot face to face
When I have crost the bar.

Chapter 9
Scarce Worth Discovery

August 4, 1864: A clear, dark, predawn sky. Both Jupiter and Saturn had set, while Mars was in the constellation of Aries, high in the south, its dark features visible through a telescope. Auriga and Taurus were both high in the east. The Moon, having just passed its new phase, did not disturb the darkness. While vacationing in Wales that summer, a young Oxford student named Gerard Manley Hopkins might have observed this scene in those hours. For all the beauty of that sky, though, Hopkins's gaze was drawn toward the second magnitude star Beta Tauri. Just west of that star shone the bright head of Tempel's comet, and its tail stretched toward another star almost as bright as the first, the nearby Iota Aurigae. The comet was moving relatively quickly, completing its quick dash past Sun and Earth within 2 weeks.

On September 13, 1864, less than a month after the speeding comet had faded from naked-eye view, Hopkins wrote these lines:

I am like a slip of comet,
Scarce worth discovery, in some corner seen
Bridging the slender difference of two stars,
Come out of space, or suddenly engender'd
By heady elements, for no man knows:
But when she sights the sun she grows and sizes
And spins her skirts out, while her central star
Shakes its cocooning mists; and so she comes
To fields of light; millions of travelling rays
Pierce her; she hangs upon the flame_cased sun,
And sucks the light as full as Gideon's fleece:
But then her tether calls her; she falls off,
And as she dwindles shreds her smock of gold
Amidst the sistering planets, till she comes
To single Saturn, last and solitary;
And then goes out into the cavernous dark.
So I go out: my little sweet is done:
I have drawn heat from this contagious sun:
To not ungentle death now forth I run. (From "Floris In Italy")

D.H. Levy, *The Starlight Night*, Astrophysics and Space Science Library 419,
DOI 10.1007/978-3-319-19878-1_9

Did Hopkins really see that predawn sky? We do not know. But even at that young age, he was a thorough scholar and a careful observer, and his earliest poems reflected his keen powers of observation. Set in the Renaissance, when "single Saturn" was indeed the outermost known planet, the poem above was intended to be a speech from an unfinished play entitled *Floris in Italy*.

The speaker, Giulia, is one of a number of people attracted to Floris, and Hopkins intended the image of the comet as Giulia's farewell soliloquy to Floris. Just as the Sun doesn't care about the planets and comets that orbit it, Floris does not recognize Giulia's worth, looking instead toward her cousin. Hopkins changed an early draft—"But when it sights the sun" to "But when she sights the sun" to strengthen the feminine metaphors of the comet image. (MacKenzie, 244–246). Like a stylish square dancer circling right or left around the square, Giulia "spins her skirts out" in a brave but unsuccessful attempt to win Floris, and as Giulia feels her distance from Floris growing, she falls off, shedding her smock of gold.

Hopkins was committed to the idea that every object in nature has its own individualizing quality. He called this special quality *inscape*; a comet, for example, has a pattern that is unique to it. This pattern not just the comet's form but its inner substance and its behavior as well. The more carefully we observe a comet, the more we perceive its inscape. The response that perception invokes in us is called *instress*. Although his comet poem appeared before Hopkins developed these intriguing ideas, its lines clearly show that Hopkins saw a special relationship between the comet's appearance and its behavior. Thus it makes sense to try to understand the comet of Hopkins's poem in relation to real comets that he might have observed, and to the rich tradition that they represent.

As we noticed in Chap. 4, the 1840s were a fertile time for comets, and the night skies in the years preceding Hopkins's poem were even more lavish. In 1858, Donati's comet swung around the Sun with a long tail curving across a large segment of the evening sky; the famous American astronomer, George Bond, stood on the terrace of Harvard's new observatory and gazed at this ghostly marvel in the night. Only 3 years later, John Tebbutt, an amateur astronomer from New South Wales, sighted a new comet that was moving rapidly northward. By July of 1861, this comet had a nucleus as bright as a first magnitude star, and had grown a protracted tail stretching over two-thirds of the visible sky. The comet caused a sensation in London's newspapers, which provided detailed reports and drawings of it. A year later, Comet Swift-Tuttle appeared with a rare spike directed toward the Sun, and a series of jets seemed to erupt from the comet's center. (This comet later became famous as the parent of the annual Perseid meteor stream; appearing each August, the stream consists of dust particles from this comet.) A full contemporary account of this comet appeared, curiously enough, in *Cornhill*, a literary magazine widely read by the intellectuals of the mid-nineteenth century.(*Cornhill*, October 1862, 550–551.) Having enjoyed the magazine since his youth, Hopkins was one of its avid readers.

On July 4, 1864, Ernst Tempel, from Marseilles, France, discovered a comet moving northeast through the constellation of Aries; an observer named Respighi independently found the comet on the next night from Bologna, Italy. Within a month it became fairly bright; on August 8, it passed only about nine million miles from Earth, about the same distance as Comet Hyakutake in 1996. I suspect that Hopkins concentrated on Comet Tempel-Respighi, which was fresh in his memory, but that he mingled recollections of all these comets when he wrote "I am like a slip of comet." Hopkins implies that at this early stage, the comet is fragile and small, "scarce worth discovery." How true these words ring. Each of the comets I discovered were faint when I first saw them, appearing as faint fuzzy patches of light, some barely brighter than the background of sky behind them and difficult to spot.

The more one studies the poem's third line "Bridging the slender difference of two stars," the more likely it seems that Hopkins had a particular comet in mind, and most likely a comet he had personally seen and observed. At first I believed the description to apply to Donati's comet of 1858, for when it was first discovered it was not far from the faint star Epsilon Leonis, moving slowly toward nearby 13 Leonis. But the line in the poem implies a comet stretching between two stars, not one appearing to move from one star to a neighboring one. Moreover, Epsilon Leonis is more than sixteen times brighter (about three magnitudes) than 13 Leonis. There is no evidence that this early path of Donati's comet reached England's popular press of the time, and Hopkins might not have known of its early journey.

At first, Comet Tempel-Respighi, since it passed from view so quickly (within a matter of weeks), seemed to be an unlikely prospect for Hopkins. It was an intrinsically small comet that became conspicuous only because it passed close to Earth. (When larger Comet Hyakutake passed as close in 1996, it was quite bright and sported a tail that stretched halfway across the sky.) The 1864 comet was brighter than second magnitude for only 2 days, during which time it matched the sky's brighter stars. Yet the London Times featured an interesting article announcing the comet in the paper:

> The comet first observed on the 5th inst (July) is now distinctly visible to the naked eye in the constellation Taurus and will become each night a more conspicuous object, its approach to the earth being very rapid…
>
> On Monday night it will be situate (sic) about five degrees to the left of the Pleiades, passing thence between the stars Iota in Auriga and Beta in Taurus, towards Theta in Gemini, near which it will probably be observed on the morning of August 7. The intensity of light towards the end of the week is from 15 to 20 times greater than during our observations this morning, so that it might be expected to attain the brightness of stars between the first and second magnitude … (*London Times*, 1 August 1864, 5e.)

When I mentioned this exciting research find to Joseph Ashbrook, then editor of *Sky and Telescope* magazine, he prepared an ephemeris that confirmed the Times' announcement that, shortly after August 1 (the "Monday night" in the extract), the comet passed between Iota Aurigae and Beta Tauri. Even though the stars' Greek letter designations suggested different brightnesses, these two particular stars had similar magnitudes. If the comet's tail followed cometary theory and pointed away from the Sun, then Hopkins might well have seen the comet and its tail "bridge the slender difference" of the two stars.

The next two lines of the poem suggest that in Renaissance thinking, the comet could have "come out of space" or have been "suddenly engender'd by heady elements" in the Earth's atmosphere. The play of which this poem was a part was set in that early time, and the idea that "no man knows" whether comets come from space would have been true then. By the Victorian age, it was common public knowledge, as seen in the many news stories that appeared at the time of the Great Comet of 1861, that comets were not formed from Earth's atmosphere. This major step forward was taken by Tycho Brahe, who demonstrated that a comet was more distant than the Moon, and later (as we have already seen) by Edmond Halley. An article in the *London Review*, appearing some four years before the poem, pointed out that the idea of comets coming from the air was proposed originally by Aristotle, and that the idea was accepted well into the Renaissance:

> Aristotle maintained that 'comets' were nothing more than 'meteors generated in the upper regions of the atmosphere.' Seneca conceived they were real stars, but that their appearance was indicative of important changes in the affairs of mankind. "For six months," he says in his book *Naturalium Questionum*, "was this comet to be seen by us, in the happy beginning of the reign of Nero…".
>
> The comet, in the estimation of the ancients, was beyond all other things in their regard, a *political star*. It was "the star of kings, emperors, and rulers;" it indicated a destiny to them, and through them was supposed to affect the condition of the people over whom they presided as sovereigns. (*London Review*, 13 July 1861, 44.)

This was a time when comets were known for what they meant rather than for what they were. Lucius Annaeus Seneca had a highly personal stake in trying to attribute comets to the good graces of his patron Nero: He was Nero's tutor when, at age 17, Nero became emperor of Rome. During the first years of Nero's reign, Seneca had considerable power in the government, but when Nero murdered his mother, Agrippina, in the year 59, he coerced Seneca into excusing this heinous act. For the sake of his own survival, Seneca was desperate to interpret the summer comet of the following year as a good omen for Rome's ruler. Seneca spent the last years of his life with his friend Burrus, trying to limit the excesses of Nero's madness. In the year 63, Burrus died and Seneca tried to resign from politics, but Nero, feigning respect for the scientist-philosopher, refused his resignation.

Seneca knew his end under Nero's tyranny was near, and he even abandoned his wealth to live in poverty in hopes of forestalling Nero's wrath. It didn't work. In AD 65, Nero accused Seneca of participating in a plot against him and ordered the scholar to "prepare for death." According to custom, this gave Seneca his choice of demise. Seneca elected to cut his wrist and then let himself bleed to death, and in this fashion the life of one of cometary astronomy's greatest scholars ended.

Centuries later, long after comets were known to travel in orbits about the Sun, politics claimed the life of another cometary scientist and threatened the career of Charles Messier, the first person to become a comet hunter specifically for the purpose of discovering a comet. Having discovered several comets, the French astronomer was famous as a gentleman scientist: Louis XV called him the ferret of comets, and he was supported by a pension from his friend, President Jean Baptiste de

Saron, of the Paris parliament. De Saron was a man versed in comet orbit calculation, as well as in politics. But with the onset of the French Revolution, Messier was forced to leave the observatory in Paris. Even so, during the evening of September 27, 1793, Messier found a comet in Ophiuchus. As he had done so many times before, he informed his friend de Saron, who attempted to calculate an orbit using the positions Messier supplied. The comet was visible only briefly before it sank into the evening twilight.

However, by this time, de Saron was no longer president of the Paris Parliament. Accused as an enemy of reform, he was in prison awaiting execution by "Madame Guillotine," that creature of the Revolution. It is hard to imagine how de Saron could have cared about comet orbits when he was about to forfeit his head, but he somehow managed to calculate, from his prison cell, an orbit for Messier's comet. If de Saron's orbit was correct, the comet would move closer to the Sun, then swing away and reappear in the morning sky. On December 29, Messier searched the eastern sky and found his comet close to the position de Saron had predicted for it. Messier wrote of de Saron's last success and hid his note in a newspaper, which he was able to smuggle to the prisoner. On April 20, 1794, just 3 months before the end of Robespierre's Reign of Terror, de Saron was guillotined. Although Messier survived, his pension was gone, and the acclaimed astronomer was virtually penniless.

Seventy years after Messier's friend died, Hopkins wrote his lines about the journey of a comet that brightens as it sights the Sun. Three comets Hopkins might have seen—Donati's of 1858, Tebutt's of 1861, and Swift-Tuttle of 1862—displayed fine inner condensations that resembled a central star shaking its cocooning mists. But the "central star" of the 1861 comet was described in this way by John Challis in the London Times of July 3, 1861:

> The large comet which has so suddenly made its appearance in the northern sky was seen here for the first time about 10 o'clock on the night of June 30… At midnight the tail reached to within a few degrees of the pole-star, and was at least 30 degrees long. The nucleus (actually the innermost part of the comet's head) is as bright as a star of the first magnitude. In the telescope the coma about the nucleus presents the singular appearance of four curved branches. (*London Times*, 3 July 1861, 5e.)

While the "star of the first magnitude" matches the image evoked in Hopkins's line, "cocooning mists"—possibly echoing the idea of protective silky threads spun by insect larvae—are common to bright comets and were especially brilliant in the heads of the comets of 1858, 1861, and 1862. Since the 1864 comet was a smaller object, bright only because of its proximity to the Earth, it had a less brilliant head.

"Shakes its cocooning mists" is a splendid image that aptly depicts the behavior of both Donati's comet of 1858 and Tebbutt's comet of 1861. Jets of dusty material from the heads of these comets streamed into their tails in periodic eruptions. The comet of 1858 released several gaseous envelopes at irregular intervals, but the 1861 comet shed 11 envelopes of material at the regular rate of one outburst every 2 days. The comet of 1862 also shook its cocooning mists, but in an unusual way: the material was apparently ejected towards the Sun instead of the more usual direction away from it. Amédeé Guillemin published this report:

> On August 10, 1862, M. Chacornac detected in the head of the comet the presence of a luminous aigrette, a brilliant sector directed towards the sun … New sectors disengaged themselves successively from the nucleus, and on August 26 M. Chacornac determined that between the 10th and the 26th they had succeeded each other to the number of thirteen. (Guillemin, 254–255.)

When a comet crosses the plane of the Earth's orbit, sunlight scattered from dust surrounding the comet might cause the appearance of a sunward spike. Comet Arend-Roland (known formally under a new designation system as C/1956 R1) in 1957, and Comet Levy, (C/1990 K1) as it faded during the early months of 1991, both displayed this unusual feature as their orbits intersected the orbit of the Earth.

The phrase "Gideon's fleece," in line 11, suggests that a comet absorbs large amounts of light from the Sun, while the space around it is dark, just as the biblical fleece soaks up water while the surrounding ground remains dry:

> And Gideon said unto God, If thou wilt save Israel by mine hand, as thou hast said,
>
> Behold, I will put a fleece of wool in the floor; and if the dew be on the fleece only, and it be dry upon all the Earth beside, then shall I know thou wilt save Israel by mine hand, as thou hast said. (*Judges* 6:36–37.)

Sometimes a comet rounds the Sun so quickly that its tail cannot keep up its role of pointing directly away from the Sun. As each dust particle in the tail assumes its own orbit, the comet leaves a tail of particles that curves away from the Sun, rather than pointing straight away from it. When the Comet of 1861 (C/1861 J1) first appeared in the sky over England, it was so bright that it could be seen in daylight. The comet really did appear to "hang upon the flame-cased Sun." (The most recent comet to display this feature was Comet Ikeya-Seki (C/1965 S1) in 1965; its head came within half a million kilometers of touching the surface of the Sun.) Comets also have gas tails made of ionized particles that rush out away from the Sun, powered by the pressure of solar radiation we call the solar wind. These tails are produced so rapidly that they generally point directly away from the Sun.

By invoking Gideon's fleece, could Hopkins have been speculating about how comets actually shine? In 1864 there was no consensus on whether a comet could soak up light and then re-emit it, or just reflect light from the Sun. In 1877, Guillemin wrote that more spectral analysis was needed before one could establish whether comets shine by light other than that reflected by the Sun. (Guillemin, 314.) But he does include observations made by Father Secchi, who "concluded that, during the first few days, the nucleus shone by its own light, 'perhaps,' the Father observes, 'on account of the incandescent state to which the comet had been brought by its close proximity to the sun.'" Secchi was on the right track; a comet can light up as its gases absorb and then re-emit solar energy like a neon light.

After its close pass by the Earth, the 1864 comet faded. Hopkins uses "tether"—a brilliant choice of words—to describe the Sun's gravitational pull on the comet.

A comet is bound to the Sun by a gravitational tether as surely as a horse or a dog is bound by a restraining tether or leash. Unlike the horse or dog who can parade about in a circle around a central pole, however, a comet orbits in an ellipse, or a parabola, with the Sun as one of the poles. Hopkins was undoubtedly familiar with the laws of celestial mechanics that Newton first proposed in the second half of the seventeenth century. Since the comet of Hopkins's poem was bound by this gravity tether, it faded as it departed from the vicinity of the Sun and the Earth. The phrase "falls off" suggests, however, that the comet was fast receding. And as the comet receded, it shredded, as Hopkins so eloquently put it, its smock of gold. As the Earth passed through the tail of the Comet of 1861, the brilliant comet seemed gold-tinged. This color is rare in a comet, but it has been observed on the other occasion during which the Earth is known to have passed through a comet's tail, that of Halley's comet, at its return in 1910.

The fourteenth line of Hopkins's comet poem originally read, "Between the sistering planets," suggesting an image of the fading comet wandering out into the outer solar system. (MacKenzie, 40). Hopkins substituted the word "Amidst" for "Between," as if to imagine the comet passing planet after planet, or at least the orbits of planet after planet, as it receded from the Sun. We remember that the poem is set in the Renaissance, so the speaker could not have been aware of the 1781 discovery of Uranus, the 1846 finding of Neptune, and certainly not Clyde Tombaugh's discovery of Pluto, which some astronomers worldwide deleted from the list of major planets in 2006. Thus to Giulia, the poem's speaker, Saturn would have been the farthest planet. The poem's closing lines relate the events of the comet's journey to that of a human life. Giulia is resigned to the end of a fine life and a satisfying relationship; like the comet, she has drawn heat from the Sun.

What may have motivated Hopkins to use the comet analogy in a play set in Renaissance Italy? Besides his fresh memory of the comet of 1864, two earlier items in the contemporary newspapers may have suggested it, one connecting the 1861 comet to Italy and the other invoking the Renaissance:

> Have we not seen quite recently, in 1861, when the great comet of that year appeared, how it was currently reported in Italy, and doubtless elsewhere, that the new star was a sign of the speedy return of Francis II and his restoration to the throne of the Two Sicilies; and also that it presaged the fall of the temporal power and the death of Pope Pius IX? "We ought not to be astonished at the persistence of these superstitions, which only the spread of science can annihilate for ever". (Guillemin, 32)

The second newspaper item concerned a debate at the Academy of Sciences about the "terrifying" comet of 1556 (C/1556 D1), which is said to have been a cause of the abdication of Charles V of Spain. With the possible exception of Genghis Khan, Charles was the Hapsburg Emperor who ruled perhaps more territories than anyone else. He retired of his own will around 1556, living quietly on his estate and indulging his interest in clocks, and died two years later. Charles's comet appeared in the southern sky of 1556, moving through Corvus and Virgo. In 1751, scientist Richard Dunthorne suggested that it was the same comet "that created astonishment throughout Europe" in 1264 (C/1264 N1). (R. Hind, 1) Dunthorne went on to prophesy a return in 1848; in 1857, with Charles's comet still unseen,

J. Russell Hind, at that time one of England's best known astronomers, refined that forecast to August of either 1858 or 1860.

Even in Hind's day, astronomers understood that the gravitational pulls of the planets can affect the period of a comet as it travels about the Sun. The astronomer wrote in 1859:

> If a comet experienced no resistance while performing its journey round that luminary, it would make its appearance after equal intervals of time … But the movements of comets are greatly disturbed by the attraction of the various planets belonging to the solar system, particularly by Jupiter and Saturn, which far exceed the rest in magnitude. (Hind, 14)

Both the *Times* of July 4, 1861, and the *Illustrated London News*, 2 days later, printed these same words:

> … the comet gave rise to an animated discussion at yesterday's sitting of the Academy of Sciences … M. Babinet remarked that M. Hind's Ephemerides of Charles V's comet gave it the precise position of the present one. (This means that by calculating the future course of the comet that appeared in 1556, it would appear in the same place that the Comet of 1861 now appears.) M. Bromine had predicted its return in 1858, and M. Hind admitted that it might return between 1856 and 1860 … If this were so, the present comet was the same that had been observed in 1556, and caused the abdication of Charles V … M. Leverrier was not of M. Babinet's opinion. M. Hind's table showed different positions which Charles V's comet might occupy in the event of its return, and the question was so undetermined that it was no wonder to find a position in the table answering to that of the present comet … (The) motion of the present comet … was so different from that given in the table, that the identity of the two comets could no longer be admitted. (*London Times*, 4 July 1861, 12; *Illustrated London News*, 6 July 1861, 3.)

The senior scientists in this debate were well known. Most of the letters to the *Times* concerning the Comet of 1861 came from Hind, who was director of the Observatory in Regent's Park, London, and had written a short book about the Comet of 1556. He began writing to the *Times* several years before the appearance of the Comet of 1861. One of these letters concerned his own prediction that the Comet of 1556 would return, and that the Comet of 1861 could well be the same. "He is sanguine himself," an article in the Journal *Titan* claimed, "that this re-appearance of the great comet of 1556 is near at hand." (Hind) As a young astronomer just beginning his career, Leverrier performed calculations which led to the discovery of Neptune in 1846. Recent investigation by amateur astronomer David A. Sargent indicates that the comets of 1264, 1556 and 1861 were not related. (D. Seargent to Levy, 11 June 2014, personal communication; see also Seargent, 99.)

In 1877, Hopkins observed Coggia's comet: "The comet," he wrote in his Journal, "I have seen it at bedtime in the west, with head to the ground, white, a soft, well-shaped tail, not big: I felt a certain awe and instress, a feeling of strangeness, flight (it hangs like a shuttlecock at the height, before it falls) …" (Hopkins, *Journals and Papers*, 13 July 1874.) The lively image of a shuttlecock implies an object that should be in motion but which, at that second, is not moving. The comet, like the shuttlecock, appeared about to plunge down toward the horizon. Used in badminton, a shuttlecock is fitted with a ring of feathers that resembles the soft, wide tail of Coggia's comet. (One reason I love this particular image is that during this period, Hopkins was possibly quite athletic, enjoying a game of badminton from time to time. And even in this sport, his mind was astute enough to notice the behavior of the shuttlecock.) Note also how Hopkins is reacting to the comet's inscape through his feeling of "awe and instress."

Not long after this comet appearance, Hopkins wrote to his mother from North Wales about a possible comet discovery. No longer at Stonyhurst, Hopkins could not discuss his observation with an astronomer there: "I have seen one three nights … in Cancer. It is small and pale but quite visible. If it is not a comet then it must be a nebula and then it is strange I should not have noticed it before … At ten o'clock it is well visible in the northeast, not high; later it would be higher." (Hopkins, *Further Letters*) This object turned out to be not a comet but the Beehive or Praesepe Cluster. Charles Messier himself included the Beehive as No. 44 in his famous list of objects that could be confused with new comets. More recently, in 1985 Tom Gehrels, while observing from Kitt Peak National Observatory in southern Arizona, confused Halley's Comet with the same cluster. (T. Gehrels to J. V. Scotti and Levy, Autumn 1985). Although Messier knew that M44 was not a comet, Hopkins should be forgiven this interesting error. When the Beehive is low in the sky, it tends to hide its true identity of a cluster of several hundred distant stars, masquerading instead as a nebulous object, or even as a comet. In December, 1878, when the letter was written, the cluster would have been rising before dawn after several months of invisibility; Hopkins might have forgotten it from the year before. In a subsequent letter he admitted his mistake: "What I took for a comet (do you remember?) turned out to be a well known nebula of great size. Praesepe it is called, in Cancer". (*Further Letters*, 137)

Hopkins's last written reference to comets appears in an 1883 letter to his friend and fellow-poet Coventry Patmore, who had included this passage in "Wedding Sermon:"

> To move
> Frantic, like comets to our bliss,
> Forgetting that we always miss,
> And so to seek and fly the sun,
> By turns, around which love should run … (Patmore, 255.)

Comets never get so close to the Sun, Patmore apparently believed, that they are swallowed by it. Hopkins was confused by Patmore's poetic style. He thought that Patmore's image was "a contrast between the long elliptic orbits of comets, with the sun almost at one end, and the short ones, practically circles, of the planets, with the sun at the center". Hopkins went on to suggest that "it might be clearer." It is true that the perihelia, or closest points to the Sun, of most comet orbits are far enough from the Sun that they do not collide. However, there is a group of comets, the Kreutz sungrazers, that might constitute a counterpoint to Patmore's idea. Sungrazing comets are exceptional because they whip around the Sun at less than one million miles from its surface. First studied in detail as a single group with a possible common origin by Heinrich Kreutz in 1888, these comets share a similar orbit. In the 1980s, the Solwind and Solar Maximum mission satellites detected a series of Kreutz sungrazers actually colliding with the Sun, and since its launch in 1995, the Solar and Heliospheric Observatory (SOHO) has discovered more than three thousand sungrazing comets, virtually all of which disintegrate when they get so close to the Sun. The grandeur of these comets depends on the time of year they round the Sun. Comet Pereyra, at perihelion in 1963, was bright but far from spectacular. But the Great Comets of 1882 and 1965 rounded the Sun not far from the September equinox, squeaking past the Sun at a small distance and putting on splendid shows. These two comets almost certainly last brushed the Sun as a single object, perhaps as the Great Comet of 1106. (Marsden, 1179.)

From his poems, letters, and journal writings, it is apparent that Hopkins's vigorous interest in observing the night sky profoundly affected his poetry. The detail of the astronomical imagery in *I am like a slip of comet* shows a conciseness and accuracy that befits a watcher of the sky. Hopkins's keen observational skill, coupled with his great imaginative power, added a special dimension to his writings and also to the tradition of the sky at night that continues to enrich the world of English poetry.

Chapter 10
Hopkins and The Starlight Night

In exploring some of Hopkins's references to the stars, what is surprising is not that Hopkins observed the stars, but the extent to which his writings portray his interest, and it is here that common literary criticism may seem inadequate. For example, in digesting "The Starlight Night", perhaps Hopkins's best known astronomical poem, at least four critics would have their readers consult the poet's Journal entry of August 17, 1877:

> As we drove home the stars came out thick: I leant hack to look at them and my heart opening more than usual praised our Lord to and in whom all that beauty comes home (1G-J254 (reference style with regard to Hopkins explained in Appendix B.)).

Gardner refers to this passage as typifying "the year in which Hopkins gazed up at the stars and praised God". Chesney, Milward, Pick and Bergonzi all quote the passage. While the poem may be associated with this astronomical occasion, it surely did not arise out of any single observing experience. The passage does, however, mark a state of development Hopkins had been approaching. As Hopkins's commitment to the Society of Jesus grew deeper his concern and love of the night sky the heavens grew more spiritual. And while this Journal entry represents his feelings I believe it has been given an undeserved prominence in an interpretation of "The Starlight Night" which underestimates Hopkins's profound love of the night sky.

Actually, "The Starlight Night" was written amidst a flurry of notes and letters on constellations, eclipses, northern lights, a possible comet, and abnormal sunsets. This was a time of intense appreciation not only of God in nature but also of nature by itself; a time of keen observation of his natural surroundings. Only 1 week after he drafted the poem, and before sending it to his mother, he recorded his disappointment at having missed an eclipse of the Moon:

D.H. Levy, *The Starlight Night*, Astrophysics and Space Science Library 419,
DOI 10.1007/978-3-319-19878-1_10

> I am much annoyed not to have seen the total eclipse of the moon last night. I saw it only when it was three parts over, the moon being dazzlingly bright and the shadow brown. Someone on the spot excused himself for not letting us know sooner by saying it was in all the almanacks. People who one wd. have thought were better informed were letting off wild remarks about disks and heavenly bodies and what not. (4B-L3-143)

One cannot blame Hopkins for this expression of annoyance. Similar experiences happen to all observers frequently. Hopkins's interest in this nearly missed eclipse may have prompted a query from his father alluding to a discussion about eclipses:

> I ought to have written before, but if I wrote it I had to write to my father about penumbras and so on and I had so much to do that I was putting it off. (7E-L3-145)

The relatively large number of allusions to the sky at the start of Hopkins's career reflects his informal interest in the universe that punctuated his Oxford days; a typical example is found in this Keatsian passage:

> You hear and, alter'd, do not hear
>
> Being a stoled apparel'd star. (4E-P121)

These lines were written only 3 months before the comet poem that was the subject of the previous chapter. Star references in the early years tend to show his keen interest in the night sky. An early Journal entry contains this description:

> Reflection of stars in water. Pointed golden drops. Gold tails.

In September, 1864, he wrote this sequence of similes:

> Stars like wold tufts.
> Stars like golden bees.
> Stars like golden rowels.
> Sky peak'd with tiny flames. (12E-P138-J46)

As the years and his career advanced, references to stars from his inquiring mind tended to become more spiritual. I use this word in the sense of being "concerned with sacred or ecclesiastical persons or things as distinct from secular" (OED). A clear example is found in this line from "On the Portrait of Two beautiful Young People":

> Where lies your landmark, seamark, or soul's star? (33E-P196)

That Hopkins's star references to become more spiritual as Hopkins grew older is significant when we consider "The Starlight Night". The poem begins with a burst of exclamation marks, one in almost every line:

> Look at the stars! Look, look up at the skies!
> O look at all the firefolk sitting in the air!
> The bright boroughs, the circle-citadels there!
> Down in dim wolves the diamond delves! The elves'–eyes!

This opening offers Hopkins a way of sharing his sheer joy and enthusiasm for the beauty of the night sky. The poem closes on a note which is decidedly spiritual, which is not surprising since it was written as he was studying for an examination in moral theology at St. Beuno's. By celebrating God's role in nature, this poem might serve as a bridge between the more physical nature of his earlier astronomical references and the more spiritual later ones.

From a naturalist's point of view "The Starlight Night" is an exciting sonnet even though its images confuse a more traditional astronomer who has accepted the old imaginary figures in the sky without bothering to devise any other more modern or personal patterns. Like Tennyson, but unlike most other Victorian poets Hopkins is rarely inexact with nature, and he certainly would not have wish to be vague in a poem that praises G-d's role in the universe. Since "Starlight Night" was written in late February (of 1877) one might expect to find its asterisms, some of them of Hopkins's own creation, in the late winter evening sky. The first asterism suggests "circle citadels". In the east one sees the faint semicircle of nine stars in Corona Borealis, a constellation made famous by the eruption of a nova on May 12, 1866 (the same star, a recurring nova now known as T Coronae Borealis, erupted again on February 9, 1946). Further to the west Auriga the charioteer offers a number of imposing stars in the shape of an oval. Hopkins might well have also linked up the stars in Ursa Major and bright Arcturus, in nearby Bootes the herdsman, to form yet another circle citadel. Finally there remains the popular "Heavenly G" that circles about a large area of sky and includes the bright stars Aldebaran, Capella, Procyon, Sirius, and Rigel before turning inward to complete the "G" at Betelgeuse. All these patterns fit the idea of circle citadels.

If one ignores the traditional but wholly arbitrary constellation boundaries, one might have better luck with this poem's star imagery. What, for example, did Hopkins have in mind when he saw the "bright boroughs"? In answering this intriguing question one might divide the sky not into constellations but into regions of brighter and fainter stars. Orion itself is near the center of one of the winter sky's brightest sectors, a richly lit area that includes the stars of the Heavenly G. Had Hopkins been observing in the predawn hours of late February he would have noticed yet another "bright borough" that is known as the summer triangle of Vega, Deneb and Altair, rising in the east. In any event, he likely noticed the section around the pole that includes Lynx and Draco and other regions near Ursa Major that are devoid of first magnitude stars, and the Cetus region in the west, that is almost completely unrepresented by bright stars. Inventing one's own constellations is not a frivolous task, and rarely a well-designed fundamental astronomy course might contain assignments such as this. Leslie Peltier began his stargazing career in just this manner. Rising late one night, he stepped outside to see stars he had never seen before:

> "My early morning meeting with all these unknown stars had been deliberately planned. I wanted to see these stars as a total stranger sees them. I wanted to see them as an earlier age of mankind had seen them… Just how, I wondered, would this unknown starland strike MY fancy. What weird figures, beasts, and monsters would I see through my still unbiased eyes?" (Peltier, 40–42)

In his study of the inscapes of this poem, James Cotter argues that "Hopkins presents a world of fairyland as a logical system through which man views the universe." (Cotter, 172) It is based on an original meaning of inscape, the "Greek scope for 'look', which maintains that there are two sets of images in the poem, one in the sky and the other in the farmyard, and that the "firefolk" of the sky clashes with the earthly 'farmyard scare'". Peter Milward suggests that with the word "down" in line 4 *("Down in dim woods the diamonds delves! The elves' –eyes!") he "apparently reflects his attention from heaven to earth where the woods are 'dim in the starlight'". But Milward adds: "Or it may be that the poet is speaking of heaven in terms of earth." Then what about the "diamond delves"? Could they imply the vast numbers of stars in regions such as the Orion nebula? At the center of this nebula is a magnificent quadrangle or trapezium of four stars called Theta Orionis, with two fainter stars completing a sextet. Hopkins is likely to have observed this marvelous sight, one of the most striking in the entire sky, through a telescope, almost certainly while he was at Stonyhurst in the same period of his life as this sonnet. The quickgold, or dewdrops, make the ground appear metaphorically almost as starry as the sky, particularly if soft moonlight reflects off the drops of dew. Also, there are regions of sky which look like dewdrops glistening in the moonlight, which Hopkins calls quickgold. Just as Leslie Peltier's imagined star patterns reinforced the great skywatcher's will to learn the conventional constellations, Hopkins's poetic asterisms do not mean that he is a mere beginner with little idea of the traditional mythological patterns of the stars. That Hopkins is aware of the ancient imaginary figures is apparent from his Journal entry of July 9, 1868, composed during a visit to Switzerland:

> Before sunrise looking out of window saw a noble scape of stars the plough all golden falling, Cassiopeia on end with her bright quains pointing to the right, the graceful bends of Perneus (sic) underneath her, and some great star whether Capella or not I am not sure risen over the brow of the mountain. (21E-J170)

During this summer morning observing session, Ursa Major lay directly over the northern horizon, being circumpolar at Hopkins's latitude. Cassiopeia rode high in the sky, opposite the north star from Ursa Major. Her bright "quains", in order of brightness, were probably Gamma and Beta Cassiopeiae. "Perneus" is obviously a slip of the pen for Perseus (Hopkins, a classical scholar, was thoroughly familiar with the story of Perseus and used it in the ninth line of his 1879 poem "Andromeda": "Her Perseus linger and leave her to her extremes?") The star he was not certain of was almost certainly would have been Capella, as that bright star would have been rising in the northeast. These quains are similar, poetically at least, to the circling of the Lesser Bear, the constellation of Ursa Minor circling Polaris as the "wain" of Tennyson's more mature poetry:

> "Unloved, by many a sandy bar,
> The brook shall babble down the plain,
> At noon or when the lesser wain
> Is twisting round the polar star; ..."
> (*In Memoriam* CL).

Hopkins's writings at the time of "Starlight Night" suggest a process or development to his thinking about the stars. The borough of heaven and Christ which is so central to "Starlight Night" is explained 3 years later in his Foundation notes (circa August, 1880). In words that echo the start of the Starlight sonnet, Hopkins teaches:

> To know what creation is LOOK AT THE SIZE OF THE WORLD. Speed of light: it would fly six or seven times round the earth while the clock ticks once. Yet it takes thousands of years to reach us from the Milky Way, which is made up of stars swarming together (though as far from one another as we are from some of them), running into one, and looking like a soft mist, and each of them a million times as big as the earth perhaps (the sun is about that). (6G-S238)

That Hopkins is amazed at the sheer size of the universe is obvious as he attempts to celebrate and understand the order of God in Nature. He correctly notes how light travels the equivalent, though in a straight line, of his figure of six or seven times around the earth "while the clock ticks once". If his clock ticked precisely once per second, this figure is close to the speed of light as it was then calculated by Leon Foucault in 1862, but before it was finalized by Albert Michaelson in the early twentieth century. The description of the stars "swarming together" fits his reference to "boroughs" in "The Starlight Night", since the stars of the Milky Way appear to swarm together in a series of long stretches that imitate the streets of a town or a highway. This is particularly appropriate with the Milky Way region in Cygnus, where a long dark nebula known as the "Great Rift" divides the "way" into two parallel "streets". In "The Loss of the Eurydice" Hopkins uses the image of the Milky Way again, to suggest the image of a country highway leading to a shrine:

> A starlight wender of ours would say
> The marvelous Milk was Walsingham Way.

In a letter to Bridges, Hopkins suggests that his "difficulty with the Milky Way is perhaps because you do not know the allusion; it is that in Catholic times Walsingham Way was a name for the Milky Way being supposed a fingerpost to our Lady's shrine at Walsingham." (3F-Ll-53)

In addition to hints of the Milky Way in "The Starlight Night", there also is a suggestion of the presence of Orion. Of all the constellations, Orion comes into the poetry of Hopkins in the most symbolic way. In the twenty-first stanza of "The Wreck of The Deutschland" he appears as Christ the "martyr master". Hillis Miller has explained the Orion image in terms of its weather portent and its relation to Christ. While Orion is an appropriate image, Miller's defense of its meteorological significance is incomplete. He is as loose in his phraseology as some of the classical scholars who knew no astronomy, stating that "Orion rises in the late fall and sets in the spring." (p. 511) Like any other group of non-circumpolar stars Orion rises and sets within any sidereal day of a few minutes under 24 hours. Miller could have written merely that Orion makes his first evening appearance in the late fall and his last evening appearance in spring. In fact, Orion's rising just before dawn in July (in Mediterranean latitudes) indicated the start of the summer planting season, his October midnight rising signals the harvest and his first evening appearance, the point of Miller's remark, heralded the ending of summer.

Hopkins was aware that other asterisms were seen as weather omens; he includes, in his translation of Horace's "Odimpfanum volgus et arceo", the fear of "the violent calendar/ haedus rise, Arcturus set." Haedus refers specifically to Zeta Aurigae, the westernmost "kid" but other sources refer to the "kids" as including Zeta,

Epsilon and Eta, and Capella too as the "goat" (Allen, p. 90f) The first autumn sighting of this asterism by Mediterranean mariners presaged the close of shipping before the winter storms. The early evening setting of Arcturus was seen as a similar omen. Miller then discusses the tradition "associating Orion unofficially with the martyrs of the Church", using Gregory's idea that Orion is the martyr, and he concludes that "the storms of winter are the symbols of persecution the martyrs mouse by their very sanctity", and in such a light the reference is a reasonable one (512f). Hopkins himself sees this relationship in "Miror Surgentum", one of his later creations in Latin. The poet describes Orion and then prays "O heavenly Jesus, you who dream up in your hand us men and these lofty stars, all things come from you." The passage opens in a sense of wonder similar to what he felt with the Starlight sonnet (I quote the translation in the Oxford edition):

> I wonder at Orion rising through the clear night, if even though the bright moon is close at hand and It presses more heavily on the small stars nor allows them to shine with her. Yet I marvel how this Orion grows up the sky and how it gleams with its own fire, which a force that is not its own light makes bright in the heavens … why, you would think that some winds had the power to whirl its seven star points round and round. (2I-P335-336)

In "Miror Surgentem" Christ is shown as a figure who can shine despite the contending influence of weaker forces within himself. The poem continues with a prayer for the new Moon "close at hand", enabling Eisto to suggest that the composition may have been written in December or January. If the poem was written around the new year, the full moon in Gemini would indeed have been close to, just a few degrees of sky from, Orion. Returning to Miller's connection of Orion with weather and season, one finds that he relates Orion as "bringer of storms to the dark side of the bay of thy blessing" in stanza 12 of the "Deutschland" Hopkins uses "bay" in its architectural sense but he applies it mostly to the sky. The term as used in this way is found in "The Loss of the Eurydice":

> And you were a liar, O blue March day.
> Bright sun lanced fire in the heavenly bay;
> (p. 72, ll. 21–22)

It figures twice also in his Journals:

> "in the opposite southwestern bay below the sun it was like clear oil but just as full of colour …" (J207)
>
> "Two large planets, … both nearly of an altitude but ONE size. Such counterparts that each seems the reflection of the other in opposite bays of the sky and not two distinct things." (8D-J193)

In his criticism of stanza 12 (*Poems*, p. 55), John Keating tries to help in this regard by comparing the dark side the bay to the "dark side" of the Moon, concluding with dubious success that both are beyond human knowledge. Hopkins must have understood what Keating did not, that the unlit portion of the Moon was surely not necessarily a mystery, and that the back side of the moon was not always dark.

Hopkins's last major comment about the stars appears in the Sermons and Devotional Writings, p. 198, and is based upon two signs that appear in heaven, from Revelation xii.

The first sign: … a woman clothed with the sun, and the moon under her feet, and upon her head a crown of twelve stars … The second "wonder in heaven" involved … a great red dragon, having seven heads and ten horns, and seven crowns upon his heads. And his tail drew the third part of the stars of heaven, and did cast them to the earth: … (*Revelations* xii, 1–4)

Hopkins commented in his notes:

> I understand the sun, moon, and stars to mean two things, first to compare the woman to the earth, this planet, which is clothed in sunlight, ministered to more humbly by its satellite, and graced by the beauty of the zodiac and other signs of the firmament, and then to her being adorned with God's grace, the service of material nature below, and the service of angels above. By the other sign that appeared in heaven', of the red or fiery dragon I understand the counterpageant … set up by Lucifer which reduced a third of the angels and he is said with his tail to have swept the third part of the stars down to the earth … As the woman is compared to the earth in the solar system so the dragon is to the constellation Draco, the tail of which sweeps through 120 degrees or a third of the sphere and which winds round the pole (… a star in the head of Draco, was then the polestar), … so as to symbolise how Satan tried to possess himself of the sovereignty of things,… (as also he wreathed himself in the Garden round the Tree of Knowledge): though he was foiled, cast from heaven, and left master only of the material world, by a figure the earth. (9J-S198)

This somewhat convoluted passage is based on a mixture of contemporary science and the mythology of two of the oldest constellations; Hercules and Draco. Lucifer was the Greek name for Venus when it shows in the morning sky. According to Paul Harvey, "The application of the name to Satan, the rebel angel who was hurled from heaven, arises from a mistaken interpretation of *Isiah* xiv.12, 'How art thou fallen from heaven, O Lucifer, son of the morning'." (Sir Paul Harvey, 494.)

Hopkins's prose, as is shown in this example, lacks the precision and art of much of his poetry. However, the poet uses prose to try interpret a puzzling and obscure prophecy that involves "signs" in the sky. At the time of its composition he was in the midst of a lone Retreat as a Tertian, his second novitiate lasting from September 1881, to August 1882. During this time he could not consult any secular material, including something as mundane as an astronomical ephemeris. It is true that while Alpha Draconis, in the tail of Draco, was at one time the pole star, as the Earth wobbles, or precesses, on its axis, the precessing pole approaches the stars in the head of the Dragon. This oversight—perhaps a slip in memory—is unimportant. It does not detract from a poet and theologian who is interested enough in the motions of his planet to know about the slow wobble of the earth that is called precession and which results in a changing position of the celestial poles. How many other people in Hopkins's position would be aware of this and use it in their religious work? This information adds considerable substance to his ideas.

Astronomy in the time of Gerard Manley Hopkins, particularly in the years from 1860 to 1889, was going through a process of consolidation, as if it were preparing for the upheavals of the twentieth century both in the theoretical astrophysics of Albert Einstein and in the observations of Harlow Shapley and Edwin Hubble. New types of instruments and processes were enabling astronomers to refine theories that had been accepted since Newton's time. An article in the August 3, 1861, edition of the *London Review*, for instance, takes advantage of the popular interest generated by the great comet of that year to outline the advantages of the spectroscope in solar and stellar astronomy. Partly because of this wealth of new instrumentation the later nineteenth century placed a high level of importance upon accuracy in observation.

Hopkins was an observer of intricate detail in nature. It would, therefore, have been in line with his general practice for him to base his poem "I am like a slip of comet" on Tempel's comet of 1864. Not satisfying himself with the spectacular sights of nature, Hopkins looked for details of less obvious significance, and in his Journals comments on obscure phenomena like the "ellipsoid comet cloud". Hopkins's observational ability appears profoundly in his comet poem. I have explained how the verb "shakes" might refer to comets releasing gases at periodic intervals from specific areas of the nucleus as it rotates on an axis. The fact that he used such an image in a poem means that it must have made an impression on him, and such a visual picture is far more likely to come from direct observation than from casual study. The imaginative use of "cocooning" in the same line also shows observational insight, for the web-like appearance of a tiny cocoon is an impression about which I have not read in the popular literature of the period; it is more likely that Hopkins observed the effect himself. It is possible also that he observed the golden yellow color of the sky on the night of June 30, 1861, when the earth passed through the tail of Tebutt's Comet.

Other examples throughout Hopkins's writings provide evidence that he was a sophisticated observer. His successful viewing of Venus in daylight is a feat which relatively few professional astronomers could match, or cared to match, even in our own time. He also suspects, as I discuss in Appendix B, that he saw the Zodiacal light on one occasion, and it is likely that line three of "Spelt from Sybil's Leaves" refers to his awareness of this interesting phenomenon of the solar system. Although the Zodiacal light can occasionally become quite bright, most amateur observers have not seen it, as there are manifold difficulties in identifying it.

When Hopkins writes about the stars, he discusses them as though he knows what he is writing about, and is comfortable doing it. The "broad bends" of Perseus and the "bright quains" of Cassiopeia appear almost as the poet's friends as he gives the impression of their spreading out majestically over a part of the summer sky. As an amateur astronomer in the original sense of the term the Latin *amare* to love, Hopkins surely would not fall into the pit reserved for the unfortunate astronomer in Chaucer's "The Miller's Tale", for coupled with his observational acuity is a keen insight into the theory that reinforces a good sighting. In the comet fragment Hopkins does not base his comments on an actual observation, but as a beautiful metaphor comparing a horse's tether or a dog's leash to the orbit of a comet about the Sun. In a later line of his unique poem he displays an unusually deep understanding of cometary orbital mechanics. His observations are not those of a romantic who gazes ignorantly at the stars, but those of a serious student who tries to understand what he observes.

Hopkins was a unique person whose intense interest in observing the sky affected the nature of his writing and thinking. Instead of the generalized and vaguely resolved thoughts that comprise poems like Wordsworth's "Star Gazers", we find in Hopkins's astronomical references a conciseness and precision that approaches that needed for formal astronomical observations. A study of these reference and allusions should increase our respect for his intellect, and, when we find them combined with great imaginative power, add an extra dimension to our appreciation of his poetry (Figs. 10.1, 10.2, and 10.3).

Fig. 10.1 The eclipse that Hopkins observed was a partial eclipse, but eclipses follow patterns, called *saros* cycles, that result in total eclipses like this one viewed from Russia on August 1, 2008. Photograph by the author

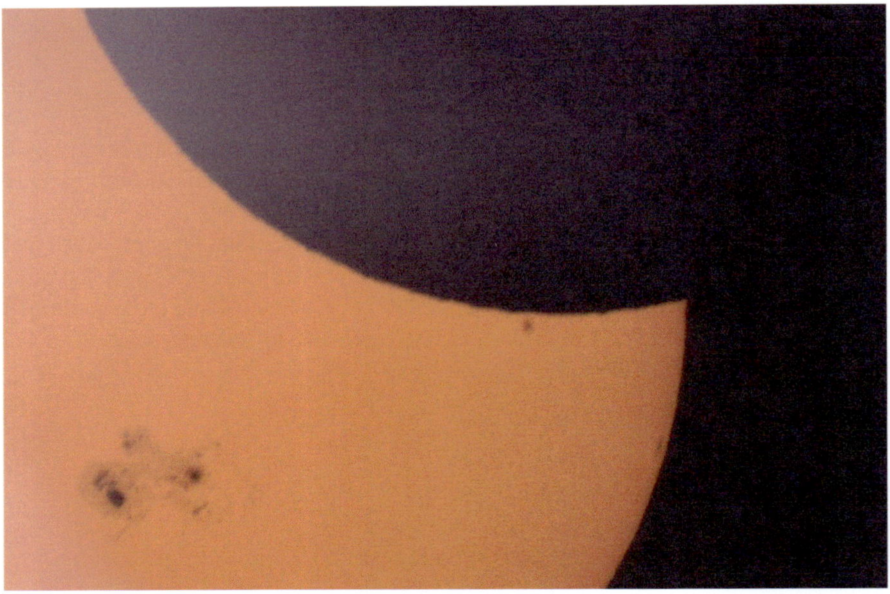

Fig. 10.2 This partial solar eclipse took place on October 23, 2014. It is similar to the eclipse that Hopkins witnessed in 1866. By a coincidence, the largest sunspot group visible within the past decade is also visible in this image. Photograph by the author

Fig. 10.3 Hopkins observed and commented on haloes around the Sun and Moon. This is an example of a lunar halo. I have seen the Moon in various unusual ways over the years. One of the most unusual was on October 27, 1971, as fog rolled in I saw the Moon to the strains of Beethoven's Moonlight sonata. Photograph by the author

Chapter 11
Conclusion

What makes both the early modern and the Victorian periods so interesting is not so much the changing theory as much as the intense nature of the observations. The supernova in Cassiopeia was the first one in over five centuries, and the brightest one in recorded history. The comet that followed only 5 years later was one of the brightest ever, and that in turn was followed by other comets and a second supernova in 1604.

Given the magnitude of these events, it stands to reason that these two events were described or alluded to in contemporary literature; for example, Henry More specifically mentioned SN 1572:

The famous Star nail'd down in Cassiopee,
How was it hammer'd in your solid sky? (More, Immortality of the Soul. 49.1–2)

… and later, a reference to both the 1572 and 1604 supernovae, and as many comets:

> Witness ye Heavens if what I say's not true,
> Ye flaming Comets wandering on high,
> And new fixt starres found in that Circle blue,
> The one espide in glittering Cassiopee,
> The other near to Ophiuchus thigh.
> Both bigger than the biggest starres that are,
> And yet as farre remov'd from mortall eye,
> As are the furthest, for those Arts declare
> Unto whose reaching sigh Heavens mysteries lie bare.
> (More, *The Infinity of Worlds*.77.1–9.)

Published in 1647, these lines can be perceived as a door that opens the world of the night sky. They emphasize that these new stars were "bigger" (meaning brighter) than all the other stars normally seen in the sky. These two passages from a work taken some decades past my major field of interest are useful because they were written by an author who could view the early modern period with a little hindsight. They are designed to stimulate discussion, with a view to open a window of

D.H. Levy, *The Starlight Night*, Astrophysics and Space Science Library 419,
DOI 10.1007/978-3-319-19878-1_11

dialogue that allows for different thoughts about the role that the new stars and comets were fulfilling in the literature of the time a few decades prior. I chose that passage because its direct reference to the star in Cassiopeia harks back to Nashe, who also referred to it directly in *Saffron Walden*, and to Jonson, who reminds his audience of the "new star" or second supernova in *Volpone* (2.1.51). The parade of new comets found voice throughout several of Shakespeare's plays, including *I Henry VI* and *Julius Caesar*. Ultimately, these references support my larger point that literature does not exist in a vacuum; it is supported and improved by its inclusion of content from other fields of endeavor. I focused on these periods and their relation to the sky, simply because some of those who wrote during those periods were familiar with that sky.

As vast and striking as our current sky remains, if we use our imaginations we can enjoy a second sky, and even a third sky, those of earlier eras whose stars, planets, and comets existed in a different context. What better choices than the sky seen during the nineteenth century, and that earlier, distant sky studied by Shakespeare and his contemporaries? The fact that Shakespeare's works contain more celestial allusions than all the other writers of the time put together might go back to his roots in rural Stratford-upon-Avon, where the night sky was unblemished by the polluted air over London; even without lights, the sky would have been diminished by dust and smoke. Especially today, the increase in light pollution is hurting our ability to appreciate the night sky we have. Going back to this earlier and simpler time allows us to value the nature of the sky that was available to the writers of the early modern era.

More than two centuries after the early modern period, astronomy in the time of Gerard Manley Hopkins, particularly in the years from 1860 to 1889, was going through a process of consolidation, as if it were preparing for the upheavals of the twentieth century both in the theoretical astrophysics of Albert Einstein and in the observations of Harlow Shapley and Edwin Hubble. New types of instruments and processes were enabling astronomers to refine theories that had been accepted since Newton's time. An article in the August 3, 1861, edition of the *London Review*, for instance, takes advantage of the popular interest generated by the great comet of that year to outline the advantages of the spectroscope in solar and stellar astronomy. Partly because of this wealth of new instrumentation the later nineteenth century placed a high level of importance upon accuracy in observation.

Hopkins was an observer of intricate detail in nature. It would, therefore, have been in line with his general practice for him to base his poem "I am like a slip of comet" on Tempel's comet of 1864. Not satisfying himself with the spectacular sights of nature, Hopkins looked for details of less obvious significance, and in his Journals comments on obscure phenomena like the "ellipsoid comet cloud". Hopkins's observational ability appears profoundly in his comet poem. I have explained how the verb "shakes" might refer to comets releasing gases at periodic intervals from specific areas of the nucleus as it rotates on an axis. The fact that he used such an image in a poem means that it must have made an impression on him, and such a visual picture is far more likely to come from direct observation than from casual study. The imaginative use of "cocooning" in the same line also shows observational insight, for the web-like appearance of a tiny cocoon is an impression about which I have not read in the popular literature of the period; it is more likely that Hopkins observed the effect himself. It is possible also that he observed the golden yellow color of the sky on the night of June 30, 1861, when the earth passed through the tail of Tebutt's Comet.

Other examples throughout Hopkins's writings provide evidence that he was a sophisticated observer. His successful viewing of Venus in daylight is a feat which relatively few professional astronomers could match, or cared to match, even in our own time. He also suspects, as I discuss in Appendix B, that he saw the Zodiacal light on one occasion, and it is likely that line three of "Spelt from Sybil's Leaves" refers to his awareness of this interesting phenomenon of the solar system. Although the Zodiacal light can occasionally become quite bright, most amateur observers have not seen it. *Sky & Telescope* Magazine, in its issue of February 1962, brings out the difficulties one has in identifying it.

When Hopkins writes about the stars, he discusses them as though he knows what he is writing about, and is comfortable doing it. The "broad bends" of Perseus and the "bright quains" of Cassiopeia appear almost as the poet's friends as he gives the impression of their spreading out majestically over a part of the summer sky. As an amateur astronomer in the original sense of the term the Latin *amare* to love, Hopkins surely would not fall into the pit reserved for the unfortunate astronomer in Chaucer's "The Miller's Tale", for coupled with his observational acuity is a keen insight into the theory that reinforces a good sighting. In the comet fragment Hopkins does not base his comments about the tether on an actual observation, but as a beautiful metaphor comparing a horse's tether or a dog's leash to the orbit of a comet about the Sun. In a later line of his unique poem he displays an unusually deep understanding of cometary orbital mechanics. His observations are not those of a romantic who gazes ignorantly at the stars, but those of a serious student who tries to understand what he observes.

Hopkins was a unique person whose intense interest in observing the sky affected the nature of his writing and thinking. Instead of the generalised and vaguely resolved thoughts that comprise poems like Wordsworth's "Star Gazers", we find in Hopkins's astronomical references a conciseness and precision that approaches that needed for formal astronomical observations. A study of these reference and allusions should increase our respect for his intellect, and, when we find them combined with great imaginative power, add an extra dimension to our appreciation of his poetry.

On a personal level, this book is intended to provide a new and different direction for people who appreciate English literature. It remains my concluding wish that this work will help inspire future students to use their knowledge of different fields to improve humanity's understanding of what these writers wished to convey, and their efforts to connect the universe in which they lived to the passion inspired by the literature they created.

Teaching literature and astronomy is not solely about teaching the canon of Shakespeare or the works of other authors, any more than it is only about teaching the cosmos. The subjects of astronomy and literature should interact with each other to inspire, inform, and critique at the same time. In the end, it is hoped that it makes readers delve more deeply into these masterworks of literature, to find parallels with other works, and develop an appreciation of what the author intended, while also seeing anew the potential for inspiration in the heavens, both as it was seen in the past and as we know it now, with the incredible array of new knowledge at the fingertips of modern science.

On the morning of April 24, 2015, I sat outdoors in my observatory to see another Lyrid meteor shower. This particular shower is not always strong, and on this night it was surely not. However, just after 3:30 a.m. my camera recorded a brilliant fireball vaporizing in the northeastern part of the sky. It was an early Eta Aquarid meteor, which means it is a piece, about the size of a small rock, that once belonged to Halley's Comet. After it separated from the comet eons ago, it orbited the Sun over and over again, until, on that fabulous morning, it disintegrated in the atmosphere of Earth, its final photons inscribing themselves on an electronic chip. Its long journey through the sky might have been interesting to a scientist, but its final flight to Earth, lasting but a second, put a poetic face on its ageless journey through the night (Fig. 11.1).

Fig. 11.1 Deep Sky night, April 24, 2015. The bright meteor at *left* is actually an early Eta Aquarid fireball, a rock-sized chunk of Halley's comet vanishing in Earth's atmosphere. Photograph by the author

Appendix A: A Selection of References to the Sky in Writings from 1572 to 1620

This appendix includes references that would normally lie beyond those chosen in the preceding chapters, but which add, in some way, to the overall goal of investigating a relationship between literature and the sky. For example, I have discovered that Shakespeare twice alludes to lunar and solar haloes, mock suns and mock moons, and uses them in astrologically predictive terms as harbingers of adversity.

Within each section the authors are presented alphabetically, except where a reference from one writer relates to that of another.

Section Codes

A	General References to Astronomy
B	Stars
C	Planets
D	Moon and Sun
E	Sunsets, Sunrises, and Night
F	Haloes
G	Meteors
H	Non-specific Astrological References

Abbreviations

Besides the standard abbreviations for Shakespeare's plays, the following abbreviations refer to specific works, as follows:

A&S *Astrophil and Stella*, Sidney
Defense *Defense of Poesie, Sidney*

D.H. Levy, *The Starlight Night*, Astrophysics and Space Science Library 419,
DOI 10.1007/978-3-319-19878-1

Malta	*The Jew of Malta*, Marlowe
OA	*Old Arcadia, Sidney*
FQ	*The Faerie Queene*, Spenser
1Tamb	*Tamburlaine, Part I*, Marlowe
2Tamb	*Tamburlaine, Part 2*, Marlowe

A. General References to Astronomy

MARLOWE

Learned Faustus,
To find the secrets of astronomy …
  Did mount him up to scale Olympus' top …
The topics zones, and quarters of the sky
  From the bright circle of the horned moon
Even to the height of primum mobile: (Faustus (circa 1592) 2.3.3242).

This chorus song occurs not long after Faustus receives his powers. Atthois level of power Fauistus' satisfaction is expressed by his relaxed smiles in Act I as he reaches from book to

Now that the gloomy shadow of the night,
Longing to view Orion's drizzling look,
  Leaps from th'antarctic world unto the sky,
  And dims the welkin with her pitchy breath, … (Faustus 1.3.14)

In a scene forecasting the play's bitter end, Faustus makes his deal under the "gloomy shadow" of darkness where even the bright stars of Orion are denied him.

That time may cease and midnight never come: (Faustus M5.2.138)

As Faustus closes, the play's action has come full circle: Faustus's knowledge and abilities are empty, and do not help to save him from damnation.

SHAKESPEARE

What zeal, what fury hath inspired thee now?
My love, her mistress, is a gracious moon;
She, an attending star, scarce seen a light …

(*LLL*.4.3.225–227)

This metaphor aligns King Ferdinand's "love, her mistress" to the bright Moon governing the night sky. In line 226 "her mistress" helps compare the King's love to the Moon, but the following line seeks to reduce her to "an attending star" that cannot even see a light, let alone emit light. Browne disputes this seeming dichotomy, insisting that "but for my love, day would turn to night!" (4.3.229) As in *I Henry VI*, where the heavens are hung with black, and as in *Macbeth*, where night strangled the lamp, this turning is sudden, as in a total eclipse.

SIDNEY

I am glad that you approve of my decision to give up the study of astronomy; but as to geometry, I don't know what I ought to do.
(Philip Sidney, Letter to Hubert Languet, (4 February 1574) 23).

This reference about "giving up" astronomy is the earliest of several found in Sir Philip Sidney's letters. The date of his letter, 4 February 1574, suggests that the supernova that appeared only 2 years earlier might have caught his interest, but that its slow fading caused the author's interest also to weaken. The frequency of astronomical allusions in Sidney's writing throughout his career does indicate that he was familiar to some extent with the night sky.

So doth the astronomer look upon the stars, and, by that he seeth, set down what order nature hath taken therein. (*Defense of Poesie* (1595), 160–162)

This report of what the astronomer does, as well as its good advice to maintain good observing records, appears frequently in the Defence. This first mention opens a wide door; the astronomer looks at the stars and then records his observations. Each further reference narrows the allusion:

For some that thought this felicity to be gotten by knowledge, no knowledge to be so high or heavenly as acquaintance with the stars, gave themselves to astronomy. … But when by the balance of experience it was found that the astronomer, looking to the stars, might fall in a ditch … (Defense (c. 1581) 300–310)

Alluding to what happens to the philosopher in Chaucer's Miller's Tale, Sidney's suggestion is a comical way of criticizing "some" for believing that the simple acquisition of knowledge is sufficient to become familiar with the stars. Sidney echoes Chaucer in suggesting that Plato accused the great astronomer Thales of having tripped and fallen while looking skyward.

Models such be wood-globes of glistening skies. (A&S (1591) 91.11)

The idea of a planetarium or planisphere, in which the stars are represented on a surface, goes back to antiquity. Near the end of Astrophel and Stella, Astrophil compares his own agony to that of the heavens reduced to a model made of wood. And much later:

BURTON

There are that observe new motions of the heavens, new stars palantia sidera, comets, clouds, call them what you will, like those Medicean, Burbonian, Austrian planets, lately detected, which do not decay, but come and go, rise higher and lower, hide and show themselves amongst the fixed stars, amongst the planets, above and beneath the moon, at set times, now nearer, now farther off, together, asunder; as he that plays upon a sackbut by pulling it up and down alters his tones and tunes, do they their stations and places, though to us undiscerned; and from those motions proceed (as they conceive) diverse alterations.
(Robert Burton. Anatomy of Melancholy (publ. 1621))

Burton here explains a relation between astronomy and geometry to which Sidney alludes earlier, but he adds music to his brew as well. The "new motions" in the heavens include comets, whose orbits can be calculated from their changing positions in the sky, and clouds, which never have orbits, as well as the traditional and well-known orbits of the planets, which Kepler published in 1605.

B. Stars

CHAPMAN

But, far above the loveliest, Hero shin'd,
And stole away th' enchanted gazer's mind;
For like sea nymphs' inveigling harmony,
So was her beauty to the standers by;
Nor that night-wandering, pale, and watery star
(When yawning dragons draw her thirling car
From Latmus' mount up to the gloomy sky,
Where, crown'd with blazing light and majesty,
She proudly sits more over-rules the flood
Than she the hearts of those that near her stood.
(Chapman, Hero and Leander, First Sestiad (circa 1611))
The "watery star" is the Moon, often visible even in a "gloomy sky" covered with thin clouds. Latmus could also imply an eclipse.

DONNE

Goe and catch a falling star … (Donne (circa 1612), Song 1)

A simple, wistful, romantic command that survives to this day in Perry Como's 1957 hit song "Catch a falling Star"; thus the theme of looking upwards on a starry evening is ancient and valid. Donne's song continues with several impossible goals, concluding that each of these is easier to attain than the love of a good woman.

FLETCHER

Seven are the lights that wander in the skies,
And at these seven I wonder in my love:
So see the moon, how pale she doth arise . . .

(Giles Fletcher, Licia (1593?) 1–3)

This sonnet opens with the number seven, a traditional celestial number since it relates to the five moving planets plus Sun and Moon. Only six objects are listed, however, and only one, the Sun, is a real star. The other five are the Moon, Mercury, Mars, Venus, and Saturn. It is possible that Fletcher means none of these in his sonnet, rather the seven stars of the Plough, or Big Dipper, or possibly the traditional "seven sister" stars of the Pleiades.

JONSON

Who heav'd Hercules Unto the Stars? or the Tyndarides?
Who placed Jasons Argo in the Sky?
Or set bright Ariadnes Crown so high?
Who made a Lamp of Berenices Hair?
Or lifted Cassiopea in her Chair?
But only Poets, rapt with Rage divine?
And such, or my hopes fail, shall make you shine.
(Jonson. Forest. 57–64 (circa 1610))

Reminiscent of the biblical story of Job, Forest demands who built the sky, and answers that divinely inspired poets did.

MARLOWE

Soldan: Nay, could their numbers countervail the stars,
or ever-drizzling drops of April showers,
Or withered leaves that autumn shaketh down …

(I Tamb. (1587) 4.1.31–33).

This comparison of the numbers of soldiers to stars in the sky, or individual droplets of rain, harks back to Genesis 22:17; "… I will multiply thy seed as the stars of the heaven, and as the sand which is upon the seashore." See also Mary Sidney's Antonie above.

But stay, what star shines yonder in the east?

The lodestar of my life, if Abigail. (Malta (1590) 2.1.41–42)
A "lodestar" or guiding star is a frequent allusion in this period (see below).
That kiss again! She runs division of my lips.
What an eye she casts on me! It twinkles like a star. (Malta 4.3.131132)

The kiss reminds Barabas when his daughter was alive; but even in death her eye resembles a star.

MARY SIDNEY

Clear stars they seem'd, which did a sun unclose
(Who, hiding none, yet all did beautify),
With coronets deck'd, with violet and rose.
(Mary Sidney The Triumph of Death (1595) 25–27)

In her long poem, Mary Sidney, Countess of Pembroke (and arguably England's first established female poet), evokes a sky full of stars, appearing with majesty and beauty after sunset, and decked with flowers. It is a strong way to open a poem whose theme is the triumph of Death over life. Indirectly the passage also relates the night sky to the permanence, and thus to the ultimate triumph, of death.

Sooner moist currents of tempestuous seas
Shall waue in heauen, and the nightly troopes
Of starres shall shine within the foming waues … (Antonie (1595) 533–535).

In this 1595 account of Antony, Sidney compares the general's war with a battle in heaven, whose troops are stars shining within waves of battle. Although it is easy to see stars reflected in still water, it is virtually impossible to see starlight reflected on rolling ocean waves.

Right as some Pallace, or some stately tower,
which ouer-lookes the nighbour buildings round
In scorning wise, and to the starres vp growes,
Which in short time his owne weight ouerthrowes.
(Tragedie of Antonie 1411–1412)

Mary Sidney constructs here a Tower-of-Babel-like palace soaring to the sky but quickly collapsing under its own bulk.

SIDNEY

Those lamps of heav'nly fire to fixed notion bound,
The ever turning spheres, the ever moving ground;
What essence destiny hath …
(Sir Philip Sidney, O.A. (circa 1580) 4.44–46).

In Sidney's era, the stars are driven by theory, according to the theory of Ptolemy rather than Copernicus; the spheres are "ever turning" but each of the stellar "lamps" are "of fixed notion bound," despite the obvious change in magnitude, though not in position, of the great star of 1572.

SPENSER

Like as a ship, whose Lodestarre suddenly
Couered with clouds, her Pilot hath dismayed … (FQ (1590) 3.53.34)

Spenser's *The Faerie Queene* is rich in celestial allusions. This example compares the loss of the ship's guiding star to Britomart's apparently aimless wanderings. A ship uses a specific lodestar that, on a clear night, will assist in its navigation, but Britomart, like the star covered by clouds, lacks that anchor as she struggles on her own journey.

Lodestarre appears again at the end of Canto 6 in FQ (1596) 3.6.52.5:

She brought her forth ino the worldes view,
To be th'ensample of true loue alone,
And Lodestarre of all chaste affection
To all faire ladies …

The Lodestarre here has a broader meaning—not a guiding star for a ship but that of a nymph called Pleasure, the daughter of Cupid and Psyche (3.6.50). See also Sonnet 34.5,10.

So I whose star, that wont with her bright ray …
My Helice the lodestar of my lyfe …

Here Spenser directly assigns the role of his personal guiding star (actually a comet with her "bright ray") to Helice, whose tail points his direction.

Night was far spent, and now in Ocean deepe
Orion, flying fast from hissing snake … (FQ (1590) 2.2.46.1–2)

Hamilton (191) proposes that Orion sets in the west pursued by his slayer; they are never together in the same hemisphere. Although this appears to be the best interpretation of the line, Scorpius is not a serpent. Serpens however, rises at about the same time but to the north, also with the ability to "chase" Orion.

… like starry light
Which sparkling on the silent waves, does seeme more bright.
(FQ (1590) 2.12.78.8–9)

Spenser accurately observes that stars seem brighter when reflected, along with the accompanying general sky glow, against the silent waveless waters of a calm ocean.

The promist ayde the tempest to withstand:
Whose loftie trees yclad with summers pride,
Did spred so broad, that heauens light did hide,
Not perceble with power of any starre: (FQ (1590) 1.1.7.3–6)

Spenser paints a summer storm scene; the swaying trees and thick leaves ("summers pride") completely cut off light from "any starre" (Hamilton 31).

… And the moist daughters of huge Atlas stroue
Into the Ocean deepe to driue their weary droue. (FQ (1590) 3.1.57.8–9)

Jupiter, "high Jove," lit the sky, while the "moist daughters" of Atlas, the Hyades star cluster (Hamilton 315), are setting into the sea toward the west. Early in 1586, 4 years before initial publication of the first three books of *Faerie Queene*, Jupiter actually passed just 1°, or two moon-diameters, to the east of the Hyades star cluster.

Now when Aldeboran was mounted hie
Above the shynie Cassiopeias chaire … (FQ (1590) 1.3.16)

When both constellations are prominent on clear winter evenings, at the latitude of England, Aldeboran (Spenser's alternate spelling of Aldebaran, or Alpha Tauri) typically rises higher in the sky than does more northern Cassiopeia (Hamilton 58). The fact that Spenser chose Cassiopeia for this particular passage raises the possibility that he saw, as a young man of 20, the supernova there in 1572.

According to F.R. Johnson, Aldebaran can be "mounted hie" over Cassiopeia only during the winter months, which contradicts the summer setting of this Canto (Johnson 194). However, I consider this contradiction resolved since Aldebaran does rise high in the sky late in August, towards the end of summer.

SHAKESPEARE

Pistol: Sweet knight, I kiss thy neaf. What! We have seen the seven stars. (2HV 2.4.176)

The "seven stars" are the Pleiades, No. 45 in Charles Messier's catalog, although seven other bright stars form the Plough or Big Dipper. The rising of the Pleiades on early November evenings is traditionally a harbinger of winter storms, which is likely the intended meaning here. Since the other seven stars of the Plough are circumpolar over England, neither rising nor setting, it is less likely that Pistol had them in mind.

Juliet: Come, gentle night; come, loving, black-brow'd night,
Give me my Romeo; and when he shall die,
Take him and cut him out in little stars,
And he will make the face of heaven so fine
That all the world will be in love with night,
And pay no worship to the garish sun. (RJ 3.3.21–25)

Juliet's legendary soliloquy summons the image of night in a novel and powerful way. Her comparison of Romeo to night is complex, because night and darkness are generally unhappy times. Unlike other invocations, this soliloquy turns night into a peaceful time where stars contrast with a blinding, dazzling sun. The stars of this night are little and innocent, not violent or malicious. The night is blackbrowed, or completely dark with no interference from day or twilight. This metaphor shows a meticulously crafted Romeo whose dead body is cut up into a pattern of little stars. At the Democratic National Convention in the Summer of 1964, Senator Robert F. Kennedy recited these lines in memory of his assassinated brother, President John Kennedy. Incidentally, these particular lines rest for eternity on the Moon's south pole, along with some ashes of the geologist E. M. Shoemaker, as a tribute to his work (Levy 262–266).

C. Planets

GRIFFIN and DOWLAND

See also the 1596 sonnet

Until it did approach my sun too near,
And then, alas, untimely was it blasted … (B. Griffin, Fidessa 35.10)
where, like a comet hurtling into the Sun, the writer's flower is killed. Also
But my sun's heavenly eyes
View not your weeping … (John Dowland, (c. 1600) *Lyrics* I.15.5–6)

As earlier, "my sun" reflects the writer's personal voice, instead of

Doth not the sun rise smiling
When fair at even he sets? (*Lyrics* 15.12–13)

In this passage it is the general Sun that rises "smiling" and not the writer's personified sun.

SIDNEY

Shall such high planets tend to the loss of a worm? (Sidney, *OA* 4.38)

Sidney asks if the planets of the higher spheres (i.e. Saturn and Jupiter) concern themselves with trivial earthly matters. Moreover:

Nay, to the heav'ns your just complainings send,
And stay the stars' inconstant constant rave
Till that they do unto our colours bend;
And ask the reason of that special grace
That they, which have no lives, should live so long,
And virtuous souls so soon should lose their place? (*OA*.Agelastus.4.37–42)

Even though there is no answer, Sidney here asks why stars, which are really not "alive" or sentient in a human sense, live so long while humans die almost at a whim. The ancient idea that stars possess souls derives from Origen, the influential second century CE theologian. "Plaint is the only pleasure," Sidney's ultimate point in *Eclogues* hammers through the tragic brevity of human lives, epigrammatic as the daily "light of sun"; in fact "this vaulty sky" looms large as "a stately tomb to cover him deceased."

Which to their eyes the bowls of Venus brought,
For they seemed made even of sky-metal best
(Sidney, Lamon's Tale (pub. circa 1598) 99–100)

The notion of the "bowls of Venus" being made of strong metal hints toward a prominent apparition of Venus in the evening sky during the spring and summer of 1580. Sky-metal implies either a substance of great strength or of sky-blue color.

She, lightning love, displaying Venus's skies … (A&S (1591) 63.7)

This rare image shows Astrophil, the lover of the stars, imagining the sky of Stella, the star he loves. In this line he may be imagining Venus not as a mythological goddess but as a geographical place in the sky. Venus, as a planet, has a sky which to Sidney is brighter and more beautiful even than that seen from Earth. Sidney might have been inspired by observing Venus shining so brightly during its 1580 visit to Earth's evening sky (A&S 63.7).

Sweet garden nymph, which keeps the cherry tree,
Whose fruit doth far th'Hesperian taste surpass … (*A&S* 79.5–6)

Like Spenser, as we shall see, Sidney uses Hesper to acknowledge the presence of Venus in the evening sky. Mythologically, Hesper's fruit lacks the taste of the cherry tree encountered by Aristophil.

SPENSER

Like Hesperus among the lesser lights … (*FQ* (1590) 1.7.30.4)

Spenser compares "a ladies head" to Venus, an evening star far brighter than any "lesser light" in the sky. The only star to claim that distinction is the Supernova of 1572, which rivalled Venus, the brightest object in the evening sky after sun and Moon, in 1572. These singular astral allusions follow a series of earlier ones; the first speaks of a "fearfull Doue" clearing the sky after rain (3.4.49), and a second during which the evening star, Venus or Hesper, appears in mid-heaven as an ascendancy "over Arthur's thoughts" (3.4.51, Hamilton 344); Venus, it turns out, was a bright evening object in late 1589, just prior to the publication of FQ's first three books (Hamilton x). However, Spenser breaches the tradition of separate names for Hesper as the evening Venus, and Phosphor as the morning Venus:

At last faire Hesperus in highest skie
Had spent his lampe …
Then up he rose, and clad him hastily … (FQ (1590) 1.2.6.6–8)

Spenser variously uses Hesperia, or Hesper, or Venus itself, to represent Venus, and never Phosphor.

he whyles his Lord in silvr slomber lay,
Like to the evening starre adorn'd wth deawy ray. (*FQ* 6.7.19.8–9)

Hamilton (669) annotates that Venus is summoned as it "rises out of the ocean", but Venus never rises in the evening; it must set in the ocean instead or it could rise before the start of morning twilight.

In 2.12.65.1–2,

As that faire star, the messenger of morne,
Her deawy face out of the sea doth reare:

Venus as Phosphor does rise in the predawn sky, though Spenser does not refer to it by any name other than as a harbinger of morning. Hamilton (45) notes how Spenser uses Hesperus to signal Venus as both an evening and a morning star, "both Venus; here named with obvious irony." I do not see the irony as so obvious, but there appears some confusion over the roles of Hesper and Phosphor which apparently remained unresolved until Tennyson settled the issue centuries later near the close of *In Memoriam* (cf. Chap. 8.)

D. Moon and Sun

CHAPMAN

… take nature's arm,
And pluck into his search the circuit

Of Earth and heaven, the sea's space and the spirit
Of every star … (George Chapman, *Learning* 18–21).

Chapman's philosophical poem is a metaphor that relates learning as being both fundamental and universal, to an understanding of all of nature. The circuit of Earth is different from the circuit of heaven. It includes periods of planetary rotation and revolution about the Earth in a Ptolmeic sense, as opposed to a distant sphere of fixed and possibly infinite numbers of stars. But Chapman offers more; within each one of those infinite suns is a spirit that goes beyond contemporary science, repeating the ancient belief in souls "equal in number to the stars" (cf. Plato *Timaeus*).

MARLOWE

Theridamus: Before the moon renew her borrowed light,
Doubt not, my lord and gracious sovereign, (*1Tamb.* (circa 1587) 1.1.69–71)

The Moon is used here as a clock; speaking before the King, Theridamus promises victory in battle within 2 weeks, by which time the waning Moon, shining by light reflected from the Sun, would have passed its new phase and begun waxing again.

SHAKESPEARE

Heaven stops the nose at it, and the moon winks … (Othello 4.2.78)

A winking Moon could imply a briefly passing cloud or an eclipse; if the latter, there was a fine partial lunar eclipse over London late in the evening of 24 May 1603, about a year before the play was completed circa 1604. From Spenser's *Faerie Queene* to Tennyson's *In Memoriam*, there is no shortage of examples of Venus (or Phosphor and Hesper) acting at different times as an evening and as a morning star, but because Mercury is normally too close to the Sun to observe, references to it in this sense are rare. The Moon appears frequently in Shakespeare especially in A Midsummer Night's Dream where Snout, Bottom, and Quince, while planning their play, try to find the phase of the Moon on the night it is to be performed (3.1.44–51). Since the Moon is to have a role in that performance, choosing the right date to perform the play is a significant part of the drama.

O, swear not by the moon, the inconstant moon
That monthly changes in her circled orb,
Lest thy love prove likewise variable. (*RJ.* 2.2. 109–111)

Juliet has just chastised Romeo for swearing his love "by yonder blessed moon." She represents the Moon as inconstant; Juliet claims either seriously or teasingly, that she does not trust that love because the Moon undergoes changes each night in its monthly lifetime as it circles the Earth. The Moon subtends a higher than earthly love (cf. Donne's "dull sublunary lover's love" Valediction 13) but Juliet suggests that Romeo's love will prove, as the lunar phases, "likewise variable." As the Sun rises and sets on lunar features, the Moon changes in shape and in brightness like the "variable star" that appeared in a different sphere in 1572. Not only was the supernova widely observed to have appeared suddenly, but it was also observed to vary in brightness from week to week as it faded slowly over the course of more than a year.

As a symbol of inconstancy in love, the inconstant Moon appears often in Shakespeare. Rosalyn claims "My face is but a moon, and clouded too." (*LLL* 5.2.203)

A few lines later she speaks to the King suggesting that the full Moon experienced an eclipse: "You took the moon at full; but now she's changed." (*LLL* 5.2.214)

The King responds dismissively "Yet still she is the Moon, and I the man." The total eclipse of the Moon of October 18, 1595, was partially visible from London. If the play were composed toward the end of the 1594–1595 window assigned to it by the *Oxford Companion to Shakespeare*, then this eclipse might have been the one alluded to. (*LLL* 5.2.2.215)

The King represents constancy as the Moon signifies variability.

In *The Tempest* Trinculo brushes off fears of Caliban's monster by comparing it to a ethereal image without substance, like "The man i' the moon!" (*Tempest* 3.1.135)

Finally, Desdemona begins to sense trouble with Othello as she plans "To follow still the changes of the moon with fresh suspicions" (*Oth.* 3.3.182–183).

Considering that eclipses of the Moon play a role in Othello, the "changes of the moon" that Desdemona plans to follow could include eclipses as well as phases; the "changes of the moon" blossom into Othello's "huge eclipse" of sun and moon (5.2.102–103).

SIDNEY

I said thine eyes were stars, thy breasts the milken way … (*A&S*, (1591) 5th song 10)

Stella spurns Astrophil's love in this fifth song.

SPENSER

His werie ghost assoyld from fleshly band,
"Did not, as others wont, directly fly

Vnto her rest in Pluto's grisly land,
Ne into ayre did vanich presently,
Ne changed was into a starre in sky;
But through traduction was eftsoones deriued (*FQ* (1596) 4.3.13)

Hamilton points out the four choices that Priamon's soul had dismissed after death: it did not join Pluto in the underworld, nor fly in the air or become a "star in sky" or join with another soul in a new body.

Or when the flying heuens he would affray;
For so exceeding shone his glistering ray,
That Phoebus golden face it did attaint,
As when a cloud his beames doth over-lay; (*FQ* (1590) 1.7.34.4–7)

The "golden face" of Phoebus has power even over the constellations, as constellations rise and set or circle the pole.

For now three moones have changed thrice their hew, (*FQ* (1590) 1.8.38.6)

Three Moons signify the passage of three months, and the following line, "haue been thrice hid underneath the ground", also clearly indicates the passage of time.

And huge Orion, that doth tempests still portend. (*FQ* (1596) 4.11.13.9)

The big constellation Orion, like the asterism of Capella, the goat with her three kids, has historically risen in the evening at the start of winter and was seen as a harbinger of winter storms.

… Return'd to heauen, whence she deriu'd her race;
Where she hath now an euerlasting place,
Mongst those twelue signs, which nightly we doe see
The heauens bright-shining baudricker to ench;
And is the Virgin, sixt in her degree,
And next her selfe in righteous balance hanging bee. (*FQ* (1596) 5.1.11.49)

Spenser here invokes Virgo as the sixth sign of the zodiac, and Libra, "righteous balance", follows. Hamilton (531) is correct about the "twelue signs" being the celestial zodiac, but not in comparing it to the Milky Way, which does not house most of the zodiac constellations, but which shares only two constellations, Sagittarius and Gemini, with the zodiac.

The which, more earth it were for mortall wight,
To tell the sands, or count the starres on high … (*FQ* (1596) 4.12.53.1–2)

Spenser is saying that it is easier to describe the sands on every beach, or count all the stars, than to think clearly in this crisis. These lines echo G-d's hyperbolical injunction to do the impossible—i.e. to count the number of the stars, and call them all by their names (*Genesis* 22.17, *Psalms* 147.4) (Hamilton 518).

Look how the Crowne, which Ariadne wore
  Vpon her ivory forehead that same day
  That Theseus her vnto his bridale bore
  When the bold Centaurus made that bloody fray …

(*FQ* (1596) 6.10/.13.1–4)

Hamilton (690) offers the insight that after Theseus abandoned Ariadne, "Bacchus placed her crown among the stars as Corona Borealis," the same crown that, centuries later, Hopkins would possibly note as "the circle-citadels there" in "The Starlight Night" (MacKenzie 139–140, 363–365).

… And twixt the twinkling of her eye-lids bright,
To sparke out little beams, like starres in foggie night. (*FQ* (1596) 6.11.21.8–9)

The simile hints that Spenser has seen the sky on a slightly foggy night, when stars appear and vanish quickly and repeatedly in large sections of sky.

E. Sunrises, Sunsets, and Night

DONNE

Busy old fool, unruly sun,
Why dost thou thus,
Through windows and through curtains call on us? …
I could eclipse and cloud them with a wink … (The Sunne rising 1–3, 12)

A favorite high school poem, "The sun rising" anthropomorphizes the Sun as an interfering dupe. It is, however, a valid, energetic, and passionate reference to sunrise. The metaphor operates directly in that the mighty Sun, in its greatness, dares to interfere with trivial human bedtime pursuits. On a loftier plane the Sun's greatness contrasts with human activities, but human beings can choose to "eclipse" the Sun by simply shutting their eyes.

MARLOWE

Zenocrate: As looks the Sun through Nilus flowing stream,
Or when the Morning holds him in her arms,
So looks my lordly love, fair Tamburlaine! (*1Tamb.* 3.2.5759)

Marlowe uses the Nilus waters as a lens. Unusual to discover in this pre-telescopic era, the stream waters are seen to refract light from the morning Sun, like the rebounding rays in Nashe's Choice of Valentines (200–202), casting the light across the water almost as a spectroscope would.

SHAKESPEARE

the all-cheering sun
Should in the farthest east begin to draw
The shady curtains from Aurora's bed … (*RJ* 1.1.132–134)

Two related sunrise allusions, one in *Romeo and Juliet*, the other in *Faerie Queene*, precursor entwines a "happie earth" to Tennyson's Ulysses (72) during which "It may be we shall touch the Happy Isles, …"

The grey-eyed morn smiles on the frowning night,
Check'ring the eastern clouds with streaks of light;
And fleckel'd darkness like a drunkard reels
From forth day's path and Titan's fiery wheels. (*RJ*.2.3.1–2)

In painting a picture of a partly cloudy summer morning, Montague invokes the first "sunrise" reference in *Romeo and Juliet*, describing how the "all-cheering sun" evaporates the morning dew. However, as the narrative moves forward these references darken; the "all-cheering sun" evolves to Friar Lawrences threatening "grey-eyed morn":

The Day begins to break, and night is fled
Whose pitchy mantle over-veil'd the earth,
Here sound retreat and cease our hot pursuit. (*1H6* 2.1–3)

This progressive darkening does not infect the comedies as much as the tragedies, but it does occur in *I Henry VI*, as well as in *Two Gentlemen of Verona*:

At first I did adore a twinkling star,
But now I worship a celestial sun. (TGV 2.6.9–10)

Proteus worships not "the Sun" which provides heat and light to our world, but "a celestial sun", again one of the infinite numbers of "suns" posited by Thomas Digges in his "Perfit Desription." The couplet also suggests that every one of these infinite twinkling stars is actually a sun.

The sun begins to gild the western sky,
And now it is about the very hour … (*TGV* 5.1.1–2)

Another of the many powerful sunset references in Shakespeare; "gild" (describes an ominous color change in the sky as the Sun descends).

SIDNEY

By what eclipse shall that Sun be defaced
What mine hath erst thrown down so fair a tower? (*OA* 2.Histor.95–96)

If Sidney wrote this passage in 1580, then he possibly was recalling the February 15 (O.S.) 1579 partial solar eclipse.

What sacrilege hath such a saint defaced? (*OA* 2.95–97)

This was a possible reference to the partial solar eclipse visible from London near sunset on 19 April 1585 (O.S.), or more likely a stronger partial eclipse visible during the morning hours of 21 July 1590 (O.S.). Sidney's notion is that the Sun itself is "defaced" by the Moon during a solar eclipse; it is, as anyone who has witnessed an advancing Moon obliterating sunspots can attest.

Towards Aurora's court a nymph doth dwell, … (*A&S* 37.5)

This painting of Stella dwelling in the court of dawn and sunrise turns a mundane image into something extraordinary in which Stella dwells in majesty.

Soul's joy, bend not those morning stars from me… (*A&S* (1591) 48.1)

Here Stella lives in a state of passion. Her "soul's joy" seems defined as a pre-dawn sky whose "morning stars" need to be seen as they are.

The shading woods seem now my sun to dark,
And stately hills disdain to look so low. (Sidney, Sonnet 18, 7–8)

Sidney's astronomical references include this one to the setting sun. Unlike those of Spenser and Shakespeare, and other writers of this period, Sidney's references tend to be personal: "my sun" instead of "the sun."

Thus to my sun do I yield, such looks her beams do afford me. (OA First Eclogues.133)
But now the stars with their strange course do bind
Me one to leave … (Sonnet 20, 5–6)

Sidney intends an astrological inference here; the sphere of stars binds "me" to leave a portion of his life.

"And so behind foul clouds of fair stars do lie hidden." (Sidney, *OA* 1.154)

The permanence of the stars, as separate bodies far from the atmosphere, is delicately acknowledged. It is often not obvious that the sky consists of both temporary "foul clouds" and permanent "fair stars."

If sunny beams shame heav'nly habitation … (*OA* 2.22)

Not only do the stars shine despite clouds, but also, Sidney points out, they do exist during daylight. The Sun's intrinsic brightness blocks our view only of the stars; the stars are in the sky during daylight.

Is she in rage? So is the sun in summer hot,
Yet harvest brings. Does she, alas, absent herself?
The sun is hid; his kindly shadows cumber not,

But when to give some grace she doth content herself,
O then it shines; then are the heavens distributed,
And Venus seems, to make up her, she spent herself. (*OA* 2.61–66)

Dorus speaks of the Sun as male when it is "encumbered" behind clouds, but female when it shines brightly and unencumbered. Venus as Hesperus or Hesper, hermaphroditic and able to assume either a male or female gender, follows the Sun in the evening sky. Venus was an "evening star" during the summer and fall of 1578. It is the permanence of the distant stars that apparently drives Sidney's thinking in this passage.

O if I had a ladder for the skies,
  I would climb up, and bring a pretty star
To wear upon her neck that open lies. (*OA* 2.16–18)

Preminiscient of haiku, these three lines show a Donnean passion that merges an allusion to biblical story of Jacob's ladder with the fun of comparing a star with a precious stone. The thought evolves further in:

Erona die? O heaven, (if heav'n there be)
Have all thy whirling course no small effect?
See all thy starry eyes this shame to see? (*OA* 2.59–62)

Sidney apparently blends a modern definition of heaven with an old query from judicial astrology, inquiring if the course of stars does not have some influence.

She comes, and straight therewith her shining twins do move
Their rays to me …
She comes, with light and warmth, which like Aurora prove
Of gentle force … (*A&S* (1591) i.76.1–2, 5–6)

The "shining twins" are part of the constellation of Gemini, rising near dawn in late summer; I wonder if Sidney did not actually see that pattern one late August or early September morning. Like Juliet, Stella is portrayed as the dawn and the sunrise.

O Mercury, foregoer to the evening,
O heav'nly huntress of the savage mountains,
O lovely star, entitled of the morning,
While that my voice doth fill those woeful valleys … (*OA* 4.7–10)

Mercury was visible several times during the late sixteenth century, as Sidney was building Arcadia. To list just a few favorable appearances of Mercury, the planet was easily visible in the evening sky in July 1570, August 1575, July 1576, and August 1581; favorable morning sky appearances include January and December 1570, April 1571, March 1572, March 1573, February 1574, and April 1578. (Mercury Chaser's Calculator, http://www.fourmilab.ch/images/3planets/elongation.html, December 2008)

Since thou from me (oh me) O sun didst pass.

Therefore esteeming all good blessings toys,
I joy in grief, and do detest all joys. (*OA* 4.98–100)

This final line is a self-referential pun that cannot be correct; if Klaius really disowns all joy, how can he accept a joy in grief? Once again, his personal sun has passed from him.

Now was our heav'nly vault deprived of the light
With sun's depart, and now the darkness of the night
Did light those beamy stars which greater light did dark. (*OA* 4.2.1–3)

A poetic description of twilight; Sidney notes how the night has its own surreal form of lighting. A combination of starlight and "airglow" (scattered sunlight on dust or water particles in the atmosphere) casts a dim glow over the darkened landscape.

SPENSER

After long stormes and tempests ouerblowne,
The sunne at length his ioyous face doth clere: (*FQ* 5.3.1.1–2)

As an optimistic way of starting a new canto, few meteorological phenomena are as encouraging as a clearing sky at the end of a series of "long stormes." Similarly, Forimell after her sorrows, can "tast of ioy." (5.3.1.9)

By this the drouping day-light gan to fade,
And yield his roome to sad succeeding night,
Who with her sable mantle gan top shade
The face of earth … (*FQ* (1590) 1.11.49.5–8)

Spenser's allusive description of twilight ushering a "sad" night is typical of a central theme of light versus dark and day vs. night in *Faerie Queene*.

But first him seemed fit, that wounded Knight
To visite, after this night's perilous passe, … (*FQ* (1596) 6.3.14.1–2)

After his injurious night, the wounded knight looks forward to better times as he enjoys his Lady's visit.

There by th'vncertaine glims of starry night … (*FQ* 6.8.48.1)

Vision seems hindered because the only light, besides the fire under the altar (Hamilton 680), is that from the sky.

His onely daughter, and his onely heyre …
As bright as doth the morning starre appeare
Out of the East, with flaming lockes bedight …

(*FQ* (1590) 1.11.21.1,4–6)

Una appears as bright as Venus. The "flaming lockes" could imply the Comet of 1580, whose "flaming lockes" of tail rivaled Venus in brightness. One of the comet's discoverers, Michael Moestlin of Batang, Indonesia, claimed that the comet's central coma was round and that it was brighter than Venus. (Kronk, 8)

In widest Ocean she her throne does reare
That ouer all the earth it may be seene;
As morning Sunne her beams dispredden cleare … (*FQ* (1590) 2.2.6–8)

One of Spenser's frequent allusions to sunrise over the ocean masks a tribute to Queen Elizabeth.

And her embracing said, O happie earth,
Whereon thy innocent feet doe euer tread,
Most virtuous virgin borne of heauenly berth … (FQ (1590) 1.10.9.1–3)

F. Haloes, Parhelia, and Parselenae

SHAKESPEARE

In 3 Henry 6 Edward cries out,

Dazzle mine eyes, or do I see three suns? Richard agrees:
Three glorious suns, each one a perfect sun,
Not separated with the racking clouds,
But sever'd on a pale clear-shining sky … (*3H6* 2.1.26–30)

As Richard builds an argument designed to predict a victory via a celestial event, he invokes a genuine atmospheric effect known as a parhelion, where a bright circular glow, or mock sun, brightens a portion of a solar halo or ring around the Sun. The description is consistent with the early-morning setting when ice crystals in high clouds refract light in the upper atmosphere. What is seen ranges from a circular halo to a complex of light patterns, including parhelia or mock suns which can rival the true sun in brightness. Richard goes on to attach a judicial astrological portent: "In this, the heaven figures some event."

Also as Richard explains, these effects dissipate as the Sun rises higher in the sky:

See, see they join, embrace, and seem to kiss,
As if they vow'd some league inviolable. (*3H6* 2.1.29–30)

In this interpretation of Richard's lines, this particular halo is as much a spiritual as a scientific event; just as a halo is an effect of a partially clouded sky in motion; it also presages the joining of a couple in love.

I Henry VI works from night to the end of night, as if celestial objects themselves starred in it. The "Comets branding their tails" image returns, two scenes later, a comet "of revenge" as "a prophet to the fall of all our foes!" (3.2.30–31)

My lord, they say five moons were seen to-night;
Four fixed, and the fifth did whirl about the other four in wondrous motion. (*KJ* 4.2.182–184)

Holinshed's report may have inspired Shakespeare to use it: "There were seene in the province of Yorke five moons, one in the east, the second in the west, the third in the north, the fourth in the south, and the fifth as it were set in the middest of the other, having manie stars about it." (Bullough, IV, 29) In his notes on King John, Irving Ribner (623) records that the allusion to "five moons is a type of natural phenomenon believed to herald disaster to a kingdom." This explanation may be accurate, but the metaphor is based on a complex aerial phenomenon called a lunar halo. It is rare but possible for such a halo to produce as many as four moon parselenae. The arrangement in the sky involves two "mock moons," one on either side of the real moon, plus one above and a fourth below it, all interconnected in a bright halo that forms a perfect circle around the true Moon. Using this explanation, it is possible to imagine the lunar haloes hiding a fifth Moon whirling about the other four along the "highway" of halo light, helped along by a vivid imagination. The entire allusion functions as an astrological portent.

The Moon's light pales as it rises and sets: "To pluck bright honour from the pale-faced moon" (*IH4*.1.3.202) has similar meaning, unless Shakespeare intends a lunar eclipse, during which the full Moon's reddish glow is also pale. Besides the 1595 lunar eclipse visible from London, there was a partial lunar eclipse fully visible from London on April 12 of the same year.

SIDNEY

This small light the moon bestows
Serves thy beams but to disclose … (*A&S* (1591) 4th song 19–20)

Unless the Moon is near its full phase, the light it "bestows" is small indeed, but a tiny fraction of the light emitted by the Sun.

With how sad steps, O moon, thou climbest the skies,
How silently, and with how wan a face … O moon, tell me,
Is constant love deemed there but want of wit? (*A&S* (1591) 31.1, 910)

The Moon "silently" climbs the sky in a dance that has not changed over the centuries between Sidney's time and that of Thomas Hardy, who wrote "At a Lunar Eclipse" after viewing one over London in the spring of 1903. Sidney's Astrophil inquires of the Moon about love, and about lovers there and here.

SPENSER

Thence, to the Circle of the Moone she clambe … (*FQ* (1609) 7.6.8.1)

Phaeton's climb into the heavens travels at least to the Moon's sphere, if not the higher spheres of the planets and fixed stars.

Eftsoones the sonne of Maia forth he sent
Downe to the Circle of the Moone … (*FQ* (1609) 7.6.16.1–2)

Mercury, the son of Maia (Hamilton 717), was mythologically the eldest of the Pleiades; in astronomy the third brightest star in the star cluster. She worked her way down to the Moon's sphere to meet Phaeton who worked up to it.

Upon her thigh her cemitaire was tide … (*FQ* (1596) 5.5.3.4)

In celebrating a joining of body art and poetry, this passage likens "cemitaire" (a curved scimitar to a crescent Moon. (Hamilton 558).

The trumpets sound, and they together run
With greedy rage, and with their faulchins smot… (*FQ* (1596) 5.12.29.2)

G. Meteors

BACON

We have high towers, the highest about half a mile in height, and some of them likewise set upon high mountains, so that the vantage of the hill with the tower is in the highest of them three miles at least. And these places we call the upper region, account the air between the high places and the low as a middle region. We use these towers, according to their several heights and situations, for insulation, refrigeration, conservation, and for the view of divers meteors—as winds, rain, snow, hail, and some of the fiery meteors also. And upon them in some places are dwellings of hermits, whom we visit sometimes and instruct what to observe. (Bacon, The New Atlantis, 32)

This story considers at least one of the earliest voyages to America. Proposing that science, in the early modern period, can lead to a more evolved, Utopian society, Bacon offers a collection of towers, one of whose specific uses is to observe the night sky with "the fiery meteors also." In the pre-telescopic era, meteors were a prime observing target.

CHAPMAN

The sun still gracing us; when now (the air
Inflamed with meteors) we discovered fair
The skipping Goat; the Horse's flaming mane;
Bearded and trained comets, stars in wane;
The burning sword …
And all else meteors, that did ill abode …

(Chapman, The Tears of Peace)

Chapman's superb address to the whimsy of time specifically mentions two kinds of comets, "bearded" and "trained," indicating his understanding of the different types of comets that have graced the sky over time, some of which he had the opportunity to view during his lifetime. "Stars in wane" appears to acknowledge that when a new star appears, it is always at its brightest at first, and then always begins to fade. Chapman's eloquent allusions to the constellations of Capricornus the goat and Pegasus the horse, are surrounded by meteor showers that appear to emanate from these constellations—the Capricornids and the Pegasids of early July, and the Orionids of late October (Edberg and Levy, 219).

MARLOWE

My sword struck fire from his coat of steel,
Even in Bithynia, where I took this Turk --
As when a fiery exhalation
Wraps in the bowels of a freezing cloud,
Fighting for passage, makes the welkin crack,
And casts a flash of lightning to the earth. (*I Tamb*. 4.2.40–45)

The "exhalation" is a comet or a meteor; probably a meteor as it is compared with lightning; in fact a "flash of lightning" is the only evidence offered for the existence of this particular meteor. The passage also suggests a cold time of year, as the meteor apparently struggles to pass out of a "freezing cloud." The meteor is personified as "fighting for passage" through the cloud; all we see of it is the explosive flash as it breaks apart. On the morning of 18 November 1997, I witnessed a single, brilliant flash that appeared in a sky covered with thin cirrus clouds. Because the clouds were hardly thick enough to produce lightning, I concluded that the flash was probably the result of a bolide, or exploding meteor, from the Leonid shower that was active that night.

… so shall our swords, our lances, and our shot
Fill all the air with fiery meteors:
Then when the sky shall wax as red as blood,

It shall be said I made it red myself,
To make me think of nought but blood and war. (*I Tamb*. 4.2.51–55)

Tamburlaine takes credit for producing a storm of meteors to focus his thoughts to fiercer battle. I do not recall any meteor shower turning the whole sky red, though haze, dust, volcanic ash, or a display of the aurora borealis could make the phenomenon theoretically possible.

In silence of thy solemn evening's walk,
Making the mantle of the richest night,
The moon, the planets, and the meteors, light
There angels in their crystal armors fight
A doubtful battle … (*1Tamb.* 5.1.146–150)

As I Tamburlaine closes, a quiet walk under the stars reveals light from the Moon, planets, and meteors, all camouflaging a great and losing "doubtful" battle for Egypt.

A meteor that might terrify the earth
And make it quake at every drop it drinks … (*1Tamb*. 5.1.462–463)

This particular meteor allusion is puzzling. Even at the close of a play rich in great battles, Marlowe could have no conception of the idea of a fireball so brilliant that it would "terrify the earth and make it quake". It could evoke the great falling of a star in Revelations 9.1. This possible allusion precedes by more than a century Georges Cuvier's theory of catastrophic events having shaped the course of Earth, and precedes by almost four centuries the discovery that great objects, like the comet or asteroid that struck the Earth 65 million years ago, have indeed tumbled from the sky in Earth's remote past.

… kindle heaps of exhalations,
That bring fiery meteors may presage Death and destruction to th' inhabitants!
Over my zenith hang a blazing star,
That may endure till heaven be dissolv'd … (*2Tamb*. 3.2.3–7)

These lines convey another apocalyptic suggestion, with both meteors and a comet hanging permanently over the zenith. Note also Tamburlaine's personal use of "my" zenith, (cf. Sidney) rather than "the" zenith, as if the "blazing star" is a fleecy cloud not far above the observer; the object hanging there is closer than the fixed stars, so that someone standing miles away would presumably not see the object directly overhead.

MARY SIDNEY

The Comets flaming through the scat'red clouds
With fiery beams, most like unbroaded haires: (*Antonie* 301–302)

Mary Sidney's metaphor represents the quick passages of meteors, rather than the sedate crossing of comets. The allusion is astrological, i.e. "by signes in earth, by signes in starry spheres …" (298)

SHAKESPEARE

Brutus: The exhalations, whizzing in the air,
Give so much light that I may read by them. (JC 2.1.44–45)

These lines are part of an extended metaphor where various aspects of the night sky add to the conspiracy against Caesar, much like

Look, how a bright star shooteth from the sky, (*VA* 815)

Brutus's exhalations are meteors, but there are also Calpurnia's comets:

"Calpurnia: When beggars die there are no comets seen;
The heavens themselves blaze forth the death of princes." (*JC* 2.2.30–31)

Near the close of *Henry VIII*, the porter and his man share a tale of how the man involved in a fight missed the meteor (actually a flaming brazier):

I miss'd the meteor once, and hit that woman … (H8 5.4.49–50).

H. Non-specific Astrological References

SHAKESPEARE

Caesar: But I am constant as the northern star,
Of whose true-fixed and resting quality
There is no fellow in the firmament. (*JC*. 3.1.50–52)

Ignoring the presages of the night that just ended, Caesar likens himself to the constancy of the north star's position in the sky. Although it was much in its same spot in the sky at the time of the writing of Julius Caesar as it is now, the North Star sank closer to the northern horizon in 44 BCE and was not the "north star" during the time of action set in this play. In 44 BCE it circled the pole more widely, which allowed much of the Southern Cross to be visible from the latitude of Rome, a sight quite invisible in Shakespeare's time, and in ours.

A calendar, a calendar! Look in the almanack; find out moonshine, find out moonshine.
Quin. Yes, it did shine that night.

Bot. Why, you may then leave a casement of the great chamber window, where we play, open: and the moon may shine in at the casement.
(*MSD* 2.2.44–50)

In raising the intricacies of the calendar, this exchange treads lightly upon one of the great English debates of the time: should the nation switch to the new calendar, as the rest of Europe did following Pope Gregory XIII's order in 1582, or stay with the older Julian calendar? The change, which involves a loss of 11 days, actually was not approved in England until 1752.

O brave new world, that hath such people in't! (*Tempest* 5.1.183–184)

Miranda's triumphant announcement of a brave new world climaxes her story on the island; finally enjoying vision for the first time, Miranda's first view is optimistic (in contrast to the mournful and sadistic attestation of a hanging that ends Aldous Huxley's *Brave New World*).

I never saw
The heavens so dim by day. (*TN* 3.3.55–56)

This was probably a reference to an anomalously stormy day, unless Shakespeare recalled the darkness of the solar eclipse of 14 December 1601 (O.S.), or the near-total solar eclipse of 27 February 1598 (O.S.).

To solemnize this day the glorious sun
Stays in his course and plays the alchemist,
Turning with splendour of his precious eye
At the meager cloddy earth to glittering gold,
The yearly course that brings this day about
Shall never see it but a holiday. (*KJ* 3.1.77–82)

This interruption in the action is probably a reference to the winter solstice.

To pluck bright honour from the pale-fac'd moon … (*1H4* 1.3.20) The descriptor "pale-faced" infers a Moon partially dimmed by haze near the horizon, or less likely a lunar eclipse.

Timon: O blessed, breeding sun, draw from the earth
Rotten humidity! (*TA*.4.2.1–2) Timon emerges from his cave long enough to feel a breath-stopping humid English summer day.
The sun's a thief, and with his great attraction
Robs the vast sea; the moon's an arrant thief,
And her pale fire she snatches from the sun … (*TA* 4.3.435–436)

Appendix B: A Catalog of Astronomical References in the Writings of Gerard Manley Hopkins

Each reference is made up in the following manner, for example:

2A-J143

where

2 = reference number in section, within which entriotes are arranged chronologically

A = section code (see next page)

J143 = reference is found in *The Journals and Papers of Gerard Manley Hopkins*, p. 143.

Or

lA-P147-J46

where

P147 = page 147 of Gardner and MacKenzie, *The Poems of Gerard Manley Hopkins, fourth edition.*

J46 = this reference can also be found in the *Journals*, p. 46.

The following abbreviations are used for various works by Hopkins. See the bibliography for fuller particulars.

P = *Poems*, 4th ed.

J = *Journals and Papers*

S = Sermons and Devotional Writings

L1 = Letters to Robert Bridges

L2 = Correspondence of Hopkins and Dixcn

L3 = *Further Letters*

D.H. Levy, *The Starlight Night*, Astrophysics and Space Science Library 419,
DOI 10.1007/978-3-319-19878-1

Section Codes

Code	Section
A	Comets
B	Eclipses
C	Moon
D	Planets
E	Stars
F	Milky Way
G	"The Starlight Night"
H	"Spelt from Sybil's Leaves"
I	Orion
J	Philosophical
K	Aurora Borealis
L	Halos
M	Letters to Nature
N	Astrological
P	Music of the Spheres

Comets

1A-P147-J46. September, 1864. "I am like a slip of comet." See Chap. 7.

2A_J143. July 5, 1866.

At a quarter to four in the afternoon in the N.E. an ellipsoid comet-cloud with horizontal (or slightly sloping) hair_texture, not equable, but gathered somewhat in three bands, namely the outline ribs and one in the midst but irregularly so.

This paragraph offers early evidence of Hopkins's keen powers of observation and of his lively use of words. The OED describes "ellipsoid" as "a solid of which all the plane sectors through one of the axes are ellipses, and all other sections ellipses or circles." The cloud could appear both ellipsoid and comet-like.

3A-J227-228. November 27, 1872.

Great fall of stars, identified with Biela's comet.

Full quotation and a comment are found above.

4A-J232. A fireball.

5A-J249. July 13, 1874. Coggia's Comet.

6A-L3-135-136. Hopkins suspects he has found a comet.

7A-L3-137. What he really had seen was the Praesepe, or Beehive, Cluster, also known as Messier 44, in the constellation of Cancer the Crab.

8A-L3-317. October 7, 1883.

The image I suppose to be a contrast between the long elliptic orbits of comets, with the sun almost at one end, and the short ones, practically circles, of the planets, with the sun at the centre. It might be clearer.

This is a note on a comet image in a poem by Coventry Patmore.

Eclipses

1B-L1-12. October 8, 1866.

"The sun was eclipsed today. I saw it all up the City Road, to such a pass have natural phenomena come."

Hopkins uses "pass" in the sense of juncture or predicament. The OED offers this 1581 example: "The worlde is come to this passe, that it counteth anie thing to bee lawfull which is delightfull." Perhaps Hopkins believed that we had reached the stage where eclipses generate little public interest.

Like all eclipses, this one repeats itself approximately every 18 years, with variations in how much of the Sun is covered and in where the shadow strikes the earth. This cycle of repetition is known as the *saros*. According to tables, the most recent eclipse of the saros sequence that includes Hopkins's event occurred on May 11, 1975, and was also visible in Europe. This was followed 2 weeks later by a spectacular total lunar eclipse on May 25, 1975, which was seen from North America by thousands of people, including me.

2B-J157.September 13, 1867.

There was an eclipse at night of the Moon, and some of the Fathers told me that from the golder (sic) colour she had had at first she became, at the eclipse and while it was going on, intensely silver, while the stars did not but became yellowish green. This description implies that for some reason he did not see the eclipse himself, possibly because he was in a retreat. This eclipse occurred in September, before the teaching or academic year, a favourite time for retreats to be held. Hopkins was beginning work at the Oratory.

3B-J220. May 22, 1872.

"I was at the Observatory to see an eclipse of the Moon. There happened to be a lunar rainbow, an arc of pale white light, colours scarcely if at all discernible." This was a partial eclipse, whose midpoint occurred at 22:00 Greenwich Mean Time. If Hopkins really did see a lunar rainbow opposite the Moon in the sky, then this was a very rare observation, although it is certainly possible. A verified lunar rainbow, or moonbow is a rainbow opposite the Moon in the sky. Light refracted off water droplets in Earth's atmosphere, and coming from the Moon, not the Sun, produces a moonbow. If it appeared as a ring surrounding the Moon, then he actually observed what is known as a lunar halo.

4B-L3-143. February 27, 1877. Hopkins almost missed this total lunar eclipse. Mid totality occurred at approximately 19:00 Greenwich Mean Time. This eclipse was near the centre of its saros cycle.

5B-L3-145. Manley Hopkins, the poet's father, had asked his son to explain what a penumbra is—the outer shadow of a body like the Earth and the Moon, or even a human from which a thin, partial, outer shadow surrounds the dark umbra The term is used most frequently to describe an eclipse. Hopkins was postponing his reply.

6B-L3-305. September 23, 1883. In a criticism of Patmore's "Wedding" Hopkins writes that "The moon is eclipsed by shadow (the Earth's shadow), the sun cannot be." The Sun obviously does not cast a shadow.

7B-L2-140. August 7 and 9, 1886. Discussing Shakespeare's minor mistakes, Hopkins remarked to Dixon: "There are some errors you must not make, as an eclipse at the halfmoon … but others do not matter and convention varies with regard to them. Hopkins was almost always accurate in his scientific comments.

The Moon (Except Lunar halos)

1C-J-(plate VII). The heading of a letter to his brother Arthur. The sky within the large "D" is dark—no stars, but a moon, indicating that the sketch was a scene around dawn or dusk.

2C-Pl16. "A Fragment of Anything You Like" about a sad and lonely Moon.

3C-Pl2. February 14, 1863.

The bugle moon by daylight floats
So glassy white about the sky

Another beautiful allusion to the Moon in daylight.

4C-Pl49. September, 1864.

Too late or else much, much too soon
Who first knew moonlight by the hunters' moon.

The close of a superb, Keatsian composition on early Autumn; the plight of a too-late flowering plant.

5C-J58. March, 1865. "The moon glassy", presumably a bright Moon in a hazy sky.

6C-P170. December 25, 1865.

Moonless darkness stands between.
Past, the Past, no more be seen!
But the Bethlehem star may lead me
To the sight of Him Who freed me
From the self that I have been.

"Moonless darkness" is an interesting description. It suggests a clear and starry night. I share Hopkins's and Peltier's belief that starlight nights are never completely dark. Light from the stars and planets themselves, as well as from a soft glow throughout the atmosphere of Earth, allows one to see a dimly lit landscape. At high altitudes the sky is even brighter; I have seen the Milky Way cast a shadow from a location at the shore of Bolivia's Lake Titicaca. A night such as this is an external, worldly darkness, not a darkness of ignorance, such as a cloudy night might represent.

7C-J161. February 25 and 26, 1868. Fine and very warm; at night the new moon almost on her back and Venus, very bright, a little to the left above—the old moon very visible. Feb. 26. This evening they were as opposite both very bright and the dark part of the moon remarkably clear and milky.—Fine and warm, with wind.

Ephemeris tables confirm this observation. At the time of this sighting on the evening of the 25th, the Moon was about 11° to the north of Venus. For his entry of the 26th. the Moon was much closer—about 3°—from Venus.

8C-J161. March 2, 1868. Venus—now very bright—with a watery nimbus and like a lamp, moon with a milky-blue iris. N.B. Both the edges of this blue are amber and sometimes rosy; the floor between the iris and the moon's disk inwards from the amber) from yellowish to bluish preen."

This appears to be a description of the Moon viewed through light haze or, more likely on a March night, through a light cirrus cloud. The various colors are seen as light is refracted through varying layers of the high cloud.

9C-J169. July 6, 1868. Sight of the moon seen from Basel.

10C-J218. February 23, 1872.

Another night ... I saw a brindled heaven. the moon just marked by a blue pot pushing its way through the darker cloud, ...

"Brindled" is also used in "Winter with the Gulf Stream"; "Its brindled wharves ..."

11C-P176. June 19, 187C. "Moonrise". See above.

12C-P68. May, 1877. "The Sea and the Skylark": "... Frequenting there while moon shall wear and wend." "Wear", in its meaning of waste, damage, sap strength, could refer to the waning of the Moon each month. "Wend" inferring as to go forward, to _"proceed" (an archaic meaning) might denote its journey each month through the constellations. Alternatively, "wear and wend" reflect the action of the tides, caused mostly by the gravity of the Moon.

Planets

1D-J19. "Ite domum saturae, venit Hesperus, ite capellae." "Go home, my cram-full goats, for the evening star arrives." A quotation from Virgil involving planet Venus as the brilliant evening star. Two other nearby astronomical references at the end of the tenth Eclogue of Virgil are worth citing because Hopkins was familiar with them. Line 69 mentions "Aethiop's flock under Cancer's star" while line 70 considers the "heavenly sisters". (Wickwam, 168)

2D—J147. July 10, 1867. "Morning star and peach—coloured dawn." Venus was a morning star that day, situated about 15° above the eastern horizon and not far from Capella. Jupiter would also have been visible in the western sky.

3D—J153. August 29, 1867. "Home by starlight and Jupiter, stumbling down steep dark lanes." Jupiter was near opposition that night, remaining in the sky all night at about magnitude minus 2, which is about sixteen times more brilliant than a first magnitude star.

4D-J161. February 25/26, 1868. Includes a sketch of Venus very near the crescent Moon. Cf. 7C-J161.

5D·J161. Observation of Venus and moon. See full quotation under references to the Moon.

6D-J162. March 10, 1868.

Venus is very bright; it lightens the quarter of the sky with a palpable nimbus—today at least, the sky being musky rather.

7DJ162. March 15, 1868. "Venus like an apple of light." Venus was a morning star, about 40° from the Sun. Perhaps he saw it near the horizon, or through haze. Venus cannot be detected as a disk by the naked eye, but as it was near its brightest phase, it may have been blurred by the atmosphere into a disk, especially if the sky was covered by thin cirrus clouds. His comment on the weather for that date, "Fine and summer-like", would also suggest that a haze was present to keep the temperature warm.

8D-J193. November 20, 1869.

Two large planets, the one an evening star, the other distant today from it as in the diagram, both nearly of an altitude and of one size—such counterparts that each seems the reflection of the other in opposite bays of the sky and not two distinct things.

On November 20, 1869, Venus was a bright evening star, near maximum eastern elongation, meaning that it was as high in the sky is it normally gets. Saturn and Mars were also evening stars, and those two worlds were only 8° apart. However, they were both near conjunction with the Sun; i.e. close to the Sun in the sky, which would have made their observation almost impossible. "The other distan t" planet would have been Jupiter, which was high in the sky and almost at opposition, or opposite the Sun in the sky.

9D-J217. November 28, 1871. "At 8:30 the Sun being up and shining bright Venus was very clear in the sky. Hopkins reports seeing Venus in daylight, a rare accomplishment indeed, but considering his flurry of observational insights around this time, I certainly believe his observation. Few observers, even among professional astronomers, have ever seen Venus in daylight. This feat requires keen eye-sight and a good understanding of Venus's position in the sky.

These lines from a fragment composed in 1864 indicate that Hopkins might have seen Venus in daylight at that time as well:

For only try by gazing to divide
One star by daylight from the strong blue air,
And find it will not therefore be descried
Because its place is known and charted there.

Stars

1E-P112-L3-9. September 3, 1862.

Silence holds breath upon her throne,
And the waked stars are all alone.

2E-P8. Christmas, 1862.

In shoals of bloom: as in unpeopled skies,
Save by two stars, more crowding lights arise,
And planets bud where'er we turn our mazed eyes.

Sirius and Capella are the brightest of eight stars of first magnitude or brighter. Venus was low in the evening sky on December 25, 1862. Mars, however, was high in the sky that evening.

3E-J17. early 1864. "The fields of heaven covered with eye-brights. White-diapered with stars." This metaphor may compare the starry sky to fields of Euphrasia, a common British wildflower.

4E-P121. June, 1864.

You hear, and alter'd, do not hear
Being a stoled apparel'd star.

From "A Voice from the World", ll. Ul-2.

5E-P127-J31. July, 1864. "And the thin stars tremble not." Alternatively, "And the lessen'd stars ray not." The "lessen'd" or fainter stars do not appear to twinkle as much as do the brighter stars.

6E-P146. July–August, 1864.

And I must have the centre in my heart
To spread the compass on the all—starr'd sky:
For only try by gazing to divide
One star by daylight from the strong blue air,
And find it will not therefore be descried
Because its place is known and charted there.

Later, "to turn the compass:" This is part of a flurry of stellar verse that was written before the Comet fragment. The speaker finds that even though "its place is known and charted" he cannot find a star in daylight. Therefore "love prescriptive" means nothing if love is not there. These lines are from Fragment iii of "Floris in Italy". Hopkins's fragment considers the beauty in a geometrical context. These words may have been inspired by, and may be an echo of, Milton's expression in *Paradise Lost* VII:224–228:

... in his hand
He took the golden compasses, prepared
In God's eternal store, to circumscribe
This universe, and all created things:
One foot he centered, and the other turned
Round through the vast profundity obscure ...

Carey and Fowler, in their edition of *The Poems of Milton* (London: Longman's. 1968). p· 788. note that "The notion of a divine geometrical construction stems" originally from Proverbs viii 27 ('When he prepared the heavens, I was there: when he set a compass upon the face of the depth').

7E-J37. "Reflection of stars in water.—Pointed golden drops. Gold tails." This seemingly innocuous metaphor is vital to anyone who has seen the reflection of stars in a still pond. At the time I was writing this portion of the book, in 1978 and 1979, I often saw stars as faint as the fourth magnitude—the limit of what I could see from my Montreal home—reflected in Jarnac Pond, northwest of the city. Decades later, as I complete this revision of *The Starlight Night*, I still enjoy viewing stars reflected in the glass-like pond at our astronomy retreat, held annually in the Adirondack Mountains in northern New York state.

8E-J38. 1864. "Real Parnassian only written by poets and is as impossible for others as poetry, as practically it is as hard to reach the moon as the stars, but something: very like it may be." Twentieth century events, particularly the human exploration of the Moon, have shown this analogy to be somewhat overstated and no longer apt. Parnassian is a late nineteenth century French style of poetry. Even Hopkins's later poetry, whose sprung rhythm style was also thought of as complex and difficult to understand, believed that Parnassian verse was hard to grasp.

9E-P138. Section xx. (7) 1864. "(Stars) float from the borders of the main."

10E—P138—J43. Section xxi. September, 1864.

Above
The vast of heaven stung; with brilliant stars.

11E-P138. Section XXII. September, 1864.

How looks the night? There does not miss a star.
The million sorts of unaccounted motes
Now quicken, sheathed in the yellow galaxy.
There is no parting or bare interstice
Where the stint compass of a skylark's wings
Would not put out some tiny golden centre.

"No parting suggests the "indivisible rays" of 1F-P138-J44."

12E-P138-J46. Section xxv. September, 1864.

The sky minted into golden sequins.

Stars like gold tufts.
Stars like Golden bees.
Stars like golden rowels.
Sky peak'd with tiny flames.

13E·P139. Section xxvi. September, 1864.

His gilded rowels
Now stars of blood.

Cf. 12E-P138-J46. Also J255, part of a description of Butterfield's Church of Babicombe:

Rafters (in the choir ceiling) there fluted and striped, webs between sown with bigger and smaller stars or rowels on pale sea—green ground.

I4E-P20. November (?). 1864.

In starry water—meads they drew
These drops: which be they? stars or dew?

15E-P139-J50. Section xxvii.

A star most spiritual, principal, preeminent
Of all the golden press.

Also xxviii:

While Phosphor, risen upon the shallowing dark,
In the ruddied county of the day's upbringing
Stood capital, eminant, ... ganfalon bearer
To all the starry press. --

Phosphor is Venus's ancient representation as a morning star.

16E-J66. 1865.

Sky orange, trail of Bronze—lit clouds, stars and streak of brilliant electrum underneath, but not for this, but effect of dark intensified foreground.

17E-P139. Section xxix. January 22, 1866.

The stars were packed so close that night
They seemed to press and stare
And gather in like hurdles bright
The liberties of air.

18E—P206. August 30, 1867. Last line, translation:

I am a tranquil star amid a waste of snows.

19E·P209.

Who stops his asking mood at par
The burly sea may quite format
Nor fear the violent calendar
At Haedus—rise, Arcturus-set.

This is a reference to the onset of the winter storm season, when Arcturus sets early in the evening, about the same time as Haedus rises. Haedus is an ancient Greek name for the faint star Zeta Aurigae, which is now understood to be a binary star system about 800 light years away.

2OE—P173. (?) 1868. References to "Starry spread", "stars like flash of fire" and "stars are packed so thick".

21E-J170. July 9, 1868. Cassiopeia and Perseus. Hopkins would have seen this exciting pair of constellations during the predawn hours of that summer morning. Although Cassiopeia is an obvious W-shaped figure, Perseus takes a little more imagination to visualize. Although he does not mention it, I suspect that Hopkins was aware of the Double open cluster that lies in Perseus below Cassiopeia's two faintest stars. The Double Cluster is indeed one of the most beautiful things in the sky.

22E-J181. July 22, 1868.

Above the Breithorn Antares sparkled like a bright crab—apple tingling in the wind.

This is exquisite. Antares is the reddest of the very bright stars. The "tingling" is the twinkling of a bright star that can be so intense when it is low in the sky that the star appears to shake.

23E—J181. July 25, 1868. From his Swiss Journal:

Up at two to ascend the Breithorn. Stars twiring brilliantly. Taurus up, a pale lirght stressily edging the eastern skyline, and lightning mingled with the dawn.

The OED, in defining "twire" as "to blink" or "to peep", points out that Shakespeare also used it with reference to the stars in Sonnet 28, line 12:

when sparkling stars twire not thou gild'st the even.

Shakespeare is probably not using this in the sense of "to blink", but in the sense of not shining, as on a cloudy night. Two earlier lines in Sonnet 28, 9 and 10, support this idea:

I tell the day, to please him, thou art bright
And dost him grace when clouds do blot the heaven: …

24E—L3-111. March l, 1870.

I practise at present the evangelical poverty which I soon hope to vow, but no one is ever so poor that he is not (without prejudice to all the rest of the world) owner of the skies and stars and everything wild that is to be found on the earth.

25E—P53. St. 5. 1876. "The Wreck of the Deutschland":

I kiss my hand
To the stars, lovely—asunder
Starlight, wafting him out of it;…

Asunder: apart from each other in position, direction or thought (OED). Also perhaps distant, so that he blows the kiss to the stars.

26E-P53. Stanza 6. 1876. "Stroke and a stress that stars and storms deliver."

Stars and storms represent the peace and strength of Christ respectively. Cf. 18E-P206, a tranquil star amid a waste of snows".

27E-S37. November 23, 1879.

We ought to give Hopkins some room here. Logically, one could produce writings that rival the Sermon, particularly from Shakespeare. But Hopkins's enthusiasm with his religion, that led him to embrace Catholicism and the Jesuits in the first place, remains infectious.

28E-S239.

"The sun and the stars shining glorify God.
They stand where he placed them, they move where
he hid them. "The heavens declare the glory of
God". They glorify God, but they do not know it. His terror, the lion is like his strength, the sea is like his greatness, the honey like his sweetness: they are something like him, they make him known, they tell of him, they give him glory,

but they do not know they do, they do not know him, they never can, they are brute things that· only think of food or think of nothing.

The quotation from Psalm xviii is well known among astronomers. Each year hundreds of stargazers and telescope makers assemble at Springfield, Vermont for their Stellafane convention and inscribed in the wood of their meeting place is the opening line of this same passage.

29E—S177. November 21, 1881. "For at his birth he was manifested ... by the star to the Mahi ..."

3OE-P220. Christmas, 1881. "Doce me gaudere,... Stella, tuo sole,... " Translation, p. 337: "Teach me to rejoice, ... O star, in your rays,... " 1

31E-L3-318.October 7, 1883.

'Whereby the complex heavens' etc: If I understand this at all it seems to me a thought condensed beyond what literature will bear. I suppose it to mean: all the hierarchies of the angels do freely the things they must do, as a stars play freely in necessary orbits.

This appears to be another mild criticism of Patmore's explanations of science.

32E-S258. January 5, 1885. Of the shepherds at the birth of Christ: "We have seen his star in the east. They were watching for it."

Just what "his star" was has been the subject of much recent work, much of it, seemingly, in planetarium shows presented around the world. The closest appearance of a major comet, Halley in 11 BC, is considered an unlikely candidate. No major exploding stars, or novae or supernovae, were reported in other literature of the period. There was, however, a rare "triple conjunction" of two planets in 7 BCE. On June 17, 2 BCE there was a conjunction of Jupiter and Saturn during which the two planets were so close that they appeared as a single brilliant object when they appeared nearest to each other, early in the evening under a middle eastern sky. The reason that I doubt that this conjunction was the star of Bethlehem was that the event was in the west, not the east.

33E-P196. Christmas, 1886.

From "On the Portrait of Two Beautiful Young People", stanza 5: "Where lies your land mark, seamark or soul's star?" As Hopkins's life progresses, references to stars show a progression towards spiritual ends. This later reference is a prominent example.

34E-J262. January 2, 1888.

"This morning I made the meditation on the Three Sins, with nothing to enter but loathing of my life and a barren submission to God's will. The body cannot rest when it is in pain nor the mind be at peace as long as something bitter distills in it and aches. This may be at any time and is at many: how then can it be pretended there is for those who feel this anythinfg worth calling happiness in this world? There is a happiness, hope, the anticipation of happiness hereafter: it is better than happiness, but it is not happiness now. It is as if one were dazzled by a spark or star in the dark, seeing it but not seeing by it: we want a light shed on our way and a happiiness spread over our life..."

It might be entirely inappropriate even to try to interpret the details of this passage, inscribed carefully in Hopkins's Journal just after New Year's Day 1888.

Hopkins was nearing the end of his life, and he was suffering in deep melancholy, a condition that affected him for the last years of his life. This dark period led to some of his greatest sonnets, now known as his Sonnets of Desolation:

No worst, there is none. Pitched past pitch of grief,
More pangs will, schooled at forepangs, wilder wring... .
Wretch, under a comfort serves in a whirlwind: all
Life death does end and each day dies with sleep.

Hopkins did not intend his poems to be published; in fact the first collected edition did not appear until 1918, edited by then poet-laureate Robert Bridges. If he believed that these sonnets would never see publication, it would seem obvious that he never expected his journal entries to be published in any form. The final sentence of the above quote infers the sudden appearance of a star, "seeing it but not seeing by it". This star could be a first dramatic view of a planet such as Venus, or a bright nova. But Hopkins uses it in the sense of surprise, so that I suspect he may have had a bright meteor in mind suddenly bursting into view among the stars. The most brilliant meteors I have seen have lit the landscape for an instant, like lightning, not long enough to allow me to see by their light. How different this behavior was from that in the opening lines of the portion of the book that deals with Hopkins, when in a burst of fun he climbed up and down a goal post; it is difficult to imagine that such a man would be capable of such massive shifts in demeanor. The idea that he was so capable testifies to the complexity of this great English poet.

35E-L1-273. September, 1887. Hopkins is explaining "Tom's Garland" to his friend Bridges. The lines from the poem read:

What! Country is honour enough in all us
- lordly head,
With heaven's lights high hung round, or,
mother-ground
That mammocks, mighty foot.

His explanation:

The head is the sovereign, who has no superior but God and from heaven receives his or her authority: we must then imagine this head as bare (see St. Paul much on this) and covered, so to say, only with the sun and stars, of which the crown is a symbol, which is an ornament but not a covering; it has an enormous hat or skull cap, the vault of heaven.

36E-PI05. July 26, 1888.

Manshape, that shone
Sheer off, disseveral, a star, death blots black out;
Lines 13–14 of "That Nature is a Heraclitean Fire ..."

37E-PI98. 1889.

Rafts and rafts of flake leaves light, dealt so,
painted on the air,

Hang as still as hawk or hawkmoth, as the stars
or as the angels there,
Like the thing that never knew the earth, never
off roots
Rose.

38E-P206-7. undated. Translated on p. 321.

There is a road which leads aloft through the stars, and a golden staircase which replaces day by day and night by night: setting foot on this road I shall mount up to the farthest heights of heaven and up to the glassy halls and to the glassy sea… O set foot with me on the road that leads up to the stars.

It is possible that Hopkins might have seen one or two of the earliest photographic plates that existed at the close of the nineteenth century. From then until the last years of the twentieth century, the story of the stars was written on glass plates of all sizes until the computer age replaced them with digital imagery. Astronomer Tom Gehrels's autobiography "*On the Glassy Sea: In Search of a world view*" (New York: American Institute of Physics, 1988) uses the metaphor to show the sky on a century's worth of glass plates.

39E-S245. undated. A meditation on death.

Do you love sunshine, starlight, fresh air, flow ers, fieldsports?—Despair then:

These Hopkins certainly loved.

40E-P105-106. July 16, 1888. "That Nature is a Heraclitean Fire."

Drowned. O pity and indignation! Manshape that shone
Sheer off, disseveral, a star, death blots black-out; …

A human life seems so meaningless; in life it shone, but in death it sheers off, disseveral, becomes aloof, puny compared with a star.

The Milky Way

1F-P1J8-J44. September, 1864.

Stars waving their indivisible rays. Sky fleeced with the Milky Way.

This entire passage concerns the Milky Way, for to the unaided eye the thousands of stars in it shine as the rays of a single large entity. The sky is fleeced, or "overspread" (OED) with its presence.

2F-P60. 1875. "With belled fire and the moth-soft Milky Way," The Milky Way in spring is not as bright as that of summer or early autumn, so "fleeced" would not be appropriate. It shines with a soft light which I have observed on countless occasions. In the southern hemisphere, where the brightest part of the Milky Way shines overhead in winter time, I have actually observed its light casting a shadow.

3F-L1-53. May 20, 1878. From "The Loss of the Eurydice",

11. 101-2

That a starlight-wender of ours would say
The marvelous Milk was Walsingham Way

I have noted Hopkins's assertion that in Catholic times, "Walsingham Way was a name for the Milky Way being supposed a fingerpost to our Lady's shrine at Walsingham." The shrine has been standing in Walsingham, in Norfolk, England, in one form or another, for almost a thousand years. The restored sanctuary still stands.

"The Starlight Night"

This section includes references that were particularly helpful in preparing the portion that deals with this famous poem.

lG-J254. August 17, 1854. "As we drove home the stars came out thick".

2G-P66. "God's Grandeur". February 2, 1877. Although "world" can be taken to mean either earth or universe, in this passage it probably means universe. Hopkins has us "look at the size of the world" and describes stars and the Milky Way. In the same passage "earth" refers to our planet. The "bent world" refers to Earth.

3G-p66. February 24, 1877. "The Starlight Night". See above.

4G-P32-L145. Variant of sonnet sent to his mother, 3 days after lunar eclipse.

5G-P82. July 27, 1879 (?). "Dress his days to a dexterous and starlight order." From "The Bugler's First Communion."

6G-S238. "Look at the size of the world".

7G-P182. about 1885. "(Margaret Clitheroe)" lines 46–7.

Heaven turned its starlight eyes below To the murder of Margaret Clitheroe.

References that May Pertain to "Spelt from Sibyl's Leaves"

The Sybil is an ancient legend dating at leats from Virgil's *Aeneid* VI (lines 11–12, 268–272, 539–543). And it appeared in other Roman writings. Hopkins might also have been inspired from this passage from Shakespeare's *Titus Andronicus* to compose this poem:

You are a young huntsman, Marcus. Let alone,
And come, I will get a leaf of brass
And with a gad of steel will write these words,
And lay it by. The angry northern wind
Will blow these sands like Sibyl's leaves abroad,
And where's our lesson then? (*TA*.4.1.100–105)

Shakespeare wrote about a long-lived Italian cave-dweller called the Sibyl; she who wrote prophecies on leaves and planted them outside the mouth of her cave. When the wind was strong enough, it blew the leaves away before they could be read. (*Norton Shakespeare*, 442)

IH-J199. Left over from the creation of the solar system are uncountable billions of microscopic particles that can occasionally be seen as a cone of light that rises from the horizon and follows the ecliptic. This glow is called the Zodiacal light.

What may be Hopkins's single sighting of this phenomenon is recorded in his Journal:

A day or two before May 14 the burnished or embossed forehead of sky over the sundown; of beautiful 'clear' ...He adds on the left-hand blank page: "Perhaps the zodiacal light". May is a good month to observe the zodiacal light as the ecliptic is high in the sky, so that Hopkins's conclusion is plausible.

2H-P97. 1885.

Earnest, earthless, equal, attuneable, vaulty, voluminous, ... stupendous
Evening strains to be time's vast, womb-of-all, home-of-all, hearse-of-all night.
Her fond yellow hornlight wound to the west, her wild hollow hoarlight hung to the height
Waste; her earliest stars, earlstars, stars principal, overbend us,
Fire-featuring heaven. For earth her be'ing has unbound; her dapple is at an end, ...

There are two common explanations for line 3. Leavis proposes that "hornlight" means soft moonlight and that "hoarlight" involves cold starlight. But this is evening, not quite night, and only the principal stars are out ("stars principal" is echoed in 98c (xxvii) as "A star most spiritual, principal ... "). Also, nowhere else does Hopkins refer to starlight as "cold". The other is from Schneider (4–8), who proposes that "as evening advances, both the warm yellow afterglow of sunset in the west and, overhead, the less friendly and less human, more cosmic-looking 'wild hollow' pale light in the dome of the upper sky 'waste' (i.e., fade.)" While these accounts might be interesting, I suggest that Hopkins is recalling his past observation of the zodiacal light. This interpretation would satisfy the entire phrase "wild hollow hoarlight hung to the height" since the zodiacal light conforms to two of the OED definitions of hollow, A "empty, vacant void" and "wanting soundness, solidity or substance.". When seen at all in midnorthern latitudes the light is very indistinct and I do recall seeing it with the outer cone brighter than the inner, which would again justify "hollow". Also, the zodiacal light does indeed appear "hung to the height" since its triangular shape can stretch almost to the zenith of the sky and appear hung in the sky like a tepee. This would also fit the setting of "Spelt from Sybil's Leaves"; with the early stars already shining, the time would be about 45 min after sunset when the zodiacal light first appears.

On March 5, 1884, a "brilliant appearance" of the zodiacal light was observed, "the cone of light being exceedingly well defined." G. M. Whipple of Kew Observatory reports this in the March 13, 1884, issue of *Nature*, p. 453. This is less than a year before Hopkins set to work on *Sybil's Leaves*. Hopkins may not have seen this display, but it is likely that he read the issue of *Nature* that contains the letter.

Astronomical and Related References of a Philosophical Nature

This section contains some· of Hopkins's thoughts concerning the philosophy of science which are not easily placed in the other categories.

IJ-J49. 1864. One of the books he must have read for "Greats" at Oxford is Francis Bacon's "*Novum Organon*" (sic). *Novum Organum* was published in Latin in 1620. According to Sir Paul Harvey (Oxford Companion to English Literature, fourth ed., p. 57), "it was Bacon's ambition to create a new system of philosophy, based on a right interpretation of nature, to replace that of Aristotle; the *Novum Organum* describes the method by which the renovation of knowledge was to be achieved and is thus the keystone to the whole system." Part of the nature to be reinterpreted, of course, is astronomy, as is seen in this passage:

> Idols which are fastened on the intellect by words are of two kinds; for either they are things which exist not, … or they are names of things which exist, but confused and ill defined … of the former kind are fortune, *primum mobile*, planetary orbs, element of fire, and such conceits, which have their rise from vain and false theories. And this sort of idols is more easily ejected, because, through a constant disowning, and the obsoleteness of theories, they may be rooted out.

2J-J75. 1864.

> Science need not interfere with genius: … some scientific basis of aesthetical criticism is absolutely needed:

3J-P138. Section xxiv. September, 1864.
Night's lantern
Pointed with pierced lights, and breaks of rays
Discover'd everywhere.

This is not the broad beam from the lantern in Poem 40, "The Lantern Out of Doors". Here, the whole sky is pictured as a lantern, the sides of which have tiny holes through which light, seen as stars, can pass.

4J -J83. (?) 1865.

All thought is of course an effort an unity (sic). This may be pursued analytically as in science or synthetically as in art or morality.

5J-LJ227. September, 1865. Here the problem of the separation of science and Christianity is explored. "I am amused to find how very far the advance of thought or science is fr. being on every side an encroachment on Christianity. "he writes to Baillie. "I think I see them retiring fr. old positions before it in important parts.""

6J-P32-33. Lent, 1866. "Nondum".

We see the glories of the earth
But not the hand that wrought them all:
Night to a myriad worlds gives birth,
Yet like a lighted empty hall

Where stands no host at door or hearth
Vacant creation's lamps appal.

And still th'abysses infinite
Surround the peak from which we gaze.
Deep calls to deep, and blackest night
Giddies the soul with blinding daze
That dares to cast is searching sight
On being a dread and vacant maze.

One could consider this poem a reflection of a prominent nineteenth century concern about the place of the Earth in the universe, and of G-d's role in it. "No answering voice comes from the skies" in Nondum . Hopkins's answer, to replace doubt with faith, develops by the eighth stanza and is perhaps a Tennysonian echo:

Let patience with her chastening wand
Dispel the doubt and dry the tear;

7J-S90. A sermon on God's providence, with its limitations, as shown for example in the celestial bodies.

"Therefore all the things we see are made and provided for us, the sun, moon, and other heavenly bodies to light us, warm us, and be measures to us of time; coal and rockoil for artificial light and heat; animals and vegetables for our food and clothing; rain, wind, and snow again to make these bear and yield their tribute to us; water and the juices of plants for our drink; air for our breathing; stone and timber for our lodging; metals for our tools and traffic; the songs of birds, flowers and their smells and colours, fruits and their taste for our enjoyment. And so on: search the whole world and you will find it a million-million fold contrivance of providence planned for our use and patterned for our admiration.

Yet this providence is imperfect, plainly imperfect. The sun shines too long and withers the harvest, the rain is too heavy and rots it or in floods spreading washes it away; the air and water carry in their currents the poison of disease; there are poison plants, venomous snakes and scorpions; the beasts our subjects rebel, not only the bloodthirsty tiger that slaughters yearly its thousands, but even the bull will gore and the stallion bite or strike; at night the moon sometimes has no light to give, at others the clouds darken her; she measures time most strangely and gives us reckonings most difficult to make and never exact enough; the coalpits and oilwells are full of explosions, fires, and outbreaks of sudden death, the sea of storms and wrecks, the snow has avalanches, the earth landslips; we contend with cold, want, weakness, hunger, disease, death, and often we fight a losing battle, never a triumphant one; everything is full of fault, flaw, imperfection, shortcoming; as many marks as there are of God's wisdom in providing for us so many marks there may he set against them of more being needed still, of something having made of this very providence a shattered frame and a broken web.

8J-S166. November 15, 1881.

To praise God oftener than what duty or even the becoming called for was a thing good In itself but not a duty. But to admire the stars is in itself indifferent. Both classes were indifferent in point of sanction.

9J-S198. Draco and Lucifer.

Draco is an astronomical reference to the constellation of the Dragon; Lucifer a pseudonym for Satan.

Aurora Borealis.

While the Aurora Borealis, or the northern lights, is not an astronomical phenomenon in the strict sense (since it does not deal with a body outside the Earth), it is directly derived from activity on the Sun, as particles from mass ejections from the solar corona interact in the earth's magnetic field with other particles in the Earth's upper atmosphere or ionosphere. Hopkins's descriptions of the aurora borealis are intricate, possibly because the constant and ghostly movement of the rays enthralled him. Some of his comments are so vivid that they remind me of displays I have seen.

A time of maximum sunspot activity—and maximum auroral activity—occurred around 1870. Solar activity occurs in 11 year cycles. After a minimum during which activity is relatively low or nonexistent, action ramps up sharply toward maximum, and then slowly falls toward the next minimum.

lK·J200. September 24, 1870. Hopkins's first aurora, seen immediately after his first arrival in a northern county far from bright city lights.

First saw the Northern Lights. My eye was caught by beams of light and dark very like the crown of horny rays the sun makes behind a cloud. At first I thought of silvery cloud until I saw that these were more luminous and did not dim the clearness of the stars in the Bear. They rose slightly radiating thrown out from the earthline. Then l saw soft pulses of light one after another rise and pass upwards arched in shape tut waveringly and with the arch broken. They seemed to float, not following the warp of the sphere as falling stars look to do but free though concentrical with it. This busy working of nature wholly independent of the earth and seeming to go on in a strain of time not reckoned by our reckoning of days and years but simpler and as if correcting the preoccupation of the world by being preoccupied with and appealing to and dated to the day of judgement was like a new witness to God and filled me with delightful fear.

Hopkins was one of the few who could tell an aurora from a cloud by the aurora's failure "to dim the clearness of "the stars in the Bear". Today this is a common means to confirm a doubtful display. The "horny rays" of the Sun in setting may well he alluded to in "Spelt from Sibyl's Leaves" though light from a horn lantern is also a possible interpretation.

That Hopkins is astute enough to know the difference between a cloud and an aurora is not as simple an accomplishment as it seems. When a faint aurora is low in the sky it is often hard to distinguish from a cloud, and occasionally I have had to use a green filter to make the identification. Hopkins saw this aurora on the night of September 24, 1870, and at that time of the year the constellation of the Great Bear is low in the sky at his latitude.

2K-J200. October 25, 1870.

A little before 7 in the evening a wonderful Aurora, the same that was seen at Rome (shortly after its seizure by the Italian government) and taken as a sign of God's anger. It gathered a little below the zenith, to the S.E. I think – a knot or crown, not a true circle, of dull blood-coloured hoerns and dropped long red beams down the sky on every side, each impaling its lot of stars. An hour or so later its colour was gone but there was still a pale crown in the same place: the skies were then clear and ashy and fresh with stars and there were flashes of or like sheet-lightning.

The "pale crown" is the coronal form that brightens the zenith during major aurora displays. Although Rome is at 41° North latitude, the north magnetic pole around which aurorae form is far away, in northern Canada. There is an arc around the magnetic pole over which aurorae tend to be most active. In any event, the aurora in Rome must have been a very rare and major display.

3K-L3-235.

At night northern lights beautiful but colourless, near the horizon in permanent birchbark downward streaks but shooting in streamers across the zenith and higher sky like breath misting and then being cut off from very sensitive glass…

5K-J217. November 10, 1871. A mention of observing the northern lights.

Halos of the Sun and Moon

While this appendix does not include most of Hopkins's meteorological remarks, haloes around the Sun and Moon are in a different and special category as they are formed by light coming from specific astronomical bodies to be refracted through and reflected by ice crystals in high cirrus clouds.

1L-J163. April 6. 1868.

There were both solar and lunar halos; faint: it deserves notice. I do not know how long the first was but the latter may have lasted hours. As long as the sky conditions remain approximately the same, a halo can last throughout the day or the night.

2L-J165. May 21, 1868.

Cooler: fine, then pale sweep, with faint solar halo. and in evening … in the clouds.

3L-J189. September 15, 1868. "One of these days there was a solar halo. Remember the solar halo as an illustration." This might have been intended for use in one of Hopkins's sermons.

4L·J211. June 17, 1871.

Solar halo at sunset. it looked bigger than usual, but this was perhaps an illusion. It was of course like a rainbow incomplete.

5L—J213. August 6, 1871.

Also saw (off the Isle of Man) high near the zenith and above the sun not a halo but the arc of a bow just like a rainbow unless rather smaller. It was convex to the sun. No rain was then or at any time falling … The time was towards sunset… .

It appears that Hopkins saw a circumzenithal arc. This phenomenon is somewhat rare, and is caused by ice crystals refracted through cirrus clouds and oriented horizontally. It is also called the Bravais arc. Some of us have probably seen one of these without even being aware of it. That Hopkins observed it, and recorded it in his Journal, is a prime example of his keen power of observation and the complexity of his mind; he did not miss much.

I observed a halo complex on the morning of September 29, 1966, (Log 3, session number 1650 S2—1). It included a circumzenithal arc. The top section that is convex to the sun may have been similar to what Hopkins saw. Three decades later

Fig. B.1 The author and his wife Wendee photographed this unique circumzenithal arc on a November afternoon in 1999

on November 2, 1999, Wendee Levy pointed out, to Carolyn Shoemaker and me, a striking halo complex that included several arcs. I observed and photographed a Brevais arc that afternoon (Fig. B.1).

6L-J218. February 23, 1872.

A lunar halo: I looked at it from the upstairs. library window. It was a grave grained sky, the strands rising a little from left to right. The halo was not quite round, for in the first place it was a little pulled and drawn below, by the refraction of the lower air perhaps, but what is more it fell in on the nether left hand side to rhyme the moon itself, which was not quite at full. I could not but strongly feel in my fancy the odd instress of this, the moon leaning on her side, as if fallen back, in the cheerful light floor within the ring, after with marical rightness and success tracing round her the ring the steady copy of her own outline.

Lunar halos are refracted and appear at 22° from the Moon in the sky, and occasionally a fainter outer halo at 46° is visible. Hopkins probably saw an incomplete halo which might have caused its non-circular appearance.

7L-J220. See 3B-J22O for full quotation. During a partial lunar eclipse Hopkins saw "a lunar rainbow, an arc of pale white light, colours scarcely if at all discernible." The lunar rainbow—an arc of pale white light—is more likely a portion of a lunar halo. True lunar rainbows or moonbows are possible—they are the result of light reflected off the Moon and then refracted through cirrus clouds in Earth's atmosphere. Hopkins could have seen such an effect in the east if the partially eclipsed Moon were low in the western sky.

Letters to *Nature*

Hopkins's letters to *Nature* concern strange sunsets and sky glows that are not strictly astronomical. However, they do concern events in the upper atmosphere, above most clouds, that were set in motion by the explosion of the mighty volcano, Krakatoa, in 1883. This eruption was so violent that much of the volcano disappeared. Dust particles rose high enough to encircle the entire globe for several years, producing strange glows and very dark lunar eclipses.

These letters form an interesting and unusual part of Hopkins's legacy, for *Nature* was, and is, one of the most respected scientific journals in the world, and I suspect that Hopkins may be the only prominent English poet to have published a scientific article in it.

I have included only the letters that contain observations of effects likely related to the eruption.

1M-L2-161. Appeared November 15, 1883 Letter "B:".

2M-L2-162. Appeared January 3, 1884, not as a letter but as a full article, under the title "The Remarkable Sunsets".

3M-(Ball, p. 148). October 30, 1884. This letter was discovered after the others, probably because it bore the initials "G. M. H." only and hence had not been included in the index to Volume XXX of *Nature*. The letter makes the startling and scientifically viable point that Krakatoa's "volcanic 'wrack' had become a satellite to the earth, like Saturn's rings, and was subject to phases, of which we are now witnessing a vivid one." That this conclusion is debatable is seen in the following reply.

4M-(Nature, 31 (188b), 28). This letter, which I discovered while exploring the files of Nature, has apparently not been republished since its first appearance there, and is the reply of one of the correspondents on whose sighting Hopkins had commented:

The Sky Glows

In using the word "corona" to designate the coloured glare which has accompanied the sun during the past year, I had no intention of employing it in its astronomical sense, but in its ordinary meteorological meaning—which "G. M. H." ... has overlooked—as referring to the coloured circles on cloud and haze frequently to be seen round the sun and moon, and classed by some observers with halos. By calling the circle now visible round the sun a "corona", I mean that in appearance and probable optical cause it is more like a meteorological corona than like a halo.

May I be allowed to point out a misprint in the first paragraph of my last letter,... where it should read "unusual sky phenomena"—the world (sic) universal having been printed for unusual.

T. W. Backhouse

Sunderland, November 8

5M-Ll-202. January 1, 1885. A comment on Bridges's poem, "Eros and Psyche":

One word on Psyche and volcanic sunsets. The description of the one over the Cretan Sea so closely agrees with an account I wrote in Nature, even to details which were local only, that it is very extraordinary: you did not see my letter, did you?

Astrological and Astro-mythological References

Since Hopkins believed that "astrology is astronomy, ordinary science—with an extraordinary science added", these references are included. Hopkins's expression is a positive and productive way of characterizing a field that modern astronomers reject as an illegitimate science. While they are probably correct in their interpretation in how astrology is generally practised now, it goes without writing that humanity's understanding of its cosmic heritage began, thousands of years ago, for the purposes of astrology.

lN-P126-J29-30. July 17, 1864. "The Lover's Stars".

The destined lover, whom his stars
More golden than the world of lights,
O'er passes bleak, o'er perilous bars
Of rivers, lead, thro' storms and nights,
Or if he leaves the West behind,
Or father'd by the sunder'd South,
Shall, when his star is zenith'd find
Acceptance round his mistress' mouth: ..

But in the other's horoscope
Bad Saturn with a swart aspect
Fronts Venus. -— His ill-launched hope
In unimperill'd haven is wreck'd.
His sick stars falter. More he may
Not win, if this be not enough
He meets upon Midsummer day
The stabbing coldness of rebuff.

The stanza about "Bad Saturn" indicates that Hopkins used the western or Ptolemaic astrological system, in which Saturn and Venus are natural enemies, Saturn being naturally evil and Venus being naturally good. In an alternative eastern system, however, the two planets are considered natural friends, so that their worst relationship can be only neutral, and not generally evil. See note 4N-S264 for another comment.

2N—P23. May 6–8, 1865. Sonnet ii of "The Beginning of the End" considers "the astrologic lore":

I must feed Fancy. Show me any one
That reads or holds the astrologic lore,
And I'll pretend the credit given of yore;
And let him prove my passion was berun

In the worst hour that's measured by the sun,
With such malign conjunctions as before
No influential heaven ever wore;
That no recorded devilish thing was done …

3N—P84. August 12, 1879. "Andromeda". Possibly by coincidence this poem is dated the day the Perseid meteors reach their normal maximum. Perseus, of course, plays a major role both in the poem and as the radiant constellation of this meteor shower.

Andromeda and Perseus are two actual constellations that rise high in the late summer sky. It is possible that Hopkins's mind may have turned to the mythological figures as a result of the prominence given to their constellations because of the meteor shower.

4N—S264. January 1, 1889.

The wise men were Magians, either Zoroastrians or at least of a religion or sect of philosophy (Sabaeism) in which astrology played a part. And astrology is, astronomy, ordinary science, with an extraordinary science added. This is called after them magic and there is therefore according to the Scripture a qood or 'white' magic, lawful in itself though positively or from its dangers it may be unlawful. That is, there is above all natural science a science which bridges over the gulf between human and superhuman knowledge, that is, enters a world of spirits, not departed souls but angels. And therefore natural bodies like the stars may exercise not only a natural, as by their light and weight, but also a preternatural influence on man. That they cannot determine his fate is plain from many reasons, among which I now see that those which convinced St Austin and St. Gregory are good. For a horocope is a momentary cast or determination of the whole heaven, to which according to the ancients, and we may say in truth, the earth is like a point: in this enormous, infinite, disproportion only one thing is to be considered, the aspect of the place of birth, that is the relation between that and the heavenly sphere: this alone decides the horoscope, for all differences here below, within the same aspect, so long as they make no difference in the horoscope itself, can count for nothing, any more than difference of position between the men or houses make any sensible difference in the parallelism of their shadows in the sun. So then two men born within a few seconds or minutes of one another, too few to change the horoscope and in the same street, must have the same fate: which is not the case.

But that the stars might not determine a fate but influence a man's constitution and with it his history is not inconceivable. From their great distance this is either small or at least difficult to observe: astronomy succeeds with difficulty in measuring for instance the heat shed by Sirius upon the earth: his actinism may be more considerable: these are natural influences: it would seem that the Magians professed to observe preternatural, that is angelic, influences, and did so...

This remark raises major concerns about the value and accuracy of astrology, and today almost all astronomers would reject even Hopkns's first statement, that the stars exercise more than a natural influence, ·· "as by their light and weight"—on humans. To develop the idea of the "extraordinary" science, two astrologies were developed, the "western" which dates at least as far back as Ptolemy, and the "eastern" which is believed to have originated in Babylon. The eastern system is more complicated and much older.

Hopkins states that "two men born within a few seconds or minutes of one another, too few to change the horoscope and in the same street, must have the same fate: which is not the case." This would be true if one interprets the western astrology. However, the eastern system makes much more use of the rapid apparent motion of the Moon. According to that system, in 5 s (assuming the moon moves 15° of. arc in 24 h) the Moon's travel through about l/50,000 of a sign makes a significant difference in one's chart.

In essence, Hopkins claims that while there may indeed be a preternatural influence of stars on humanity, astrology as we know it offers too simplistic a description of it.

Direct or Indirect References to "Music of the Spheres"

The concept of music and dancing of the spheres has been used to show the order in the universe. In the fifth century B. C. Pythagoras suggested that "there is a geometry in the humming of the strings. There is music in the spacings of the spheres."5 I have collected Hopkins's writings that relate to this idea:

lP-P164. August, 1865 (?)

O what a silence is this wilderness!
Might we not think the sweet (?) and daring rises
Of the flown skylark, and his traverse flight
At highest when he seems to brush the clouds,
Had been more fertile and had sown with notes
The unenduring fallows of the heaven?
Or take it thus—that the concording stars
Had let such music down, without impediment
Falling along the breakless pool of air,
As struck with rings of sound the close—shut palms
Of the wood-sorrel and all things sensitive?

2P—P216. April, 1876. The original and discarded opening to "Ad Episcopum Salopensem" is translated on p. 330 and includes:

Just so the sky repeats its rhythms in fixed regions, and "the stars you see shining there you will see again", it says. "Just if Cassiope (a constellation) is more lovely to you in a certain quarter, it is in this quarter that the lovely Cassiope awaits you".

Hopkins compares the faces on a revolving Roman clay facepot to the "rhythms" of the sky:

The faces on the urn do not change as it revolves: "see, they recur: you do not see another man who comes with a new face."

BP-S200. November 14, 1881.

It (the singing of the angels) is also compared to a concert of music, the ranks of the angelic hierarchies being like notes of a scale and a harmonic series: the working of the commonwealth and building of the tower or temple would he like the playing on these notes, like the tune, the music. They are also compared to heavenly spheres, planetary distances, and so on: and indeed these things, music and astronomy, are compared among themselves (in the Music of the Spheres and the morning stars jump for joy)...

(morning stars reflect Shakespene's *Merchant of Venice*, V.i. 60–63:
There's not the smallest orb which thous behold'st
But in his motion like an angel sings,
Still choiring the young-eyed cherubins.
Such harmony is in immortal souls, ...

Works Cited A. Chaps. 1–5

Akrigg, G. P., ed. *Letters of King James VI and I.* Berkeley: University of California Press, 1984.

Albanese, Denise. *New Science: New World.* Durham: Duke University Press, 1996.

Allen, Don Cameron. *The Star-crossed Renaissance: The Quarrel about Astrology and its Influence in England.* 1941. New York: Octagon Books, 1973.

Allen, Richard Hinckley. *Star-Names and their Meanings.* 1899. New York: Dover, 1963.

Aston, Margaret E. (1970) "The Fiery Trigon Connunction: An Elizabethan Astrological Prediction." Isis 61,2, 158–187.

Bacon, Francis. *The Advancement of Learning.* 1605: Philadelphia: Paul Dry Books, 2001.

Bacon, Francis. New Atlantis. 1626. Los Angeles: Philosophical Research Society 1985.

Bainbridge, John. *An Astronomical Description of the Late Comet from the 18 of Novemb. 1618 to the 16 of December Following.* London, 1619.

Bertram, Paul. *Shakespeare and The Two Noble Kinsmen* (New Brunswick, New Jersey: Rutgers University Press, 1965), 244. Bible World History Timeline. 23 March 2006. <Timelinehttp://agards-bibletimeline.com/bible_study_aid:12BC.

Bourne, Michael, *Inuentions or Deuices* (London: 1578, 1585), 92.

Bradley, A. C. *Shakespearean Tragedy*. 1904. 2nd ed. London: Macmillan, 1964.

Brahe, Tycho. *Epistolae*. Hven, Denmark, 1596.

Brahe, Tycho. *Astronomiae instauratae progymnasmata*. Prague, 1602.

Brahe, Tycho. *Epistolarum astronomicarum libri* (Uraniburgi: 1596).

Brahe, Tycho. *Learned Tico Brahe his Astronomicall Cometur of the new and much admired * Which appeared in the Year 1572* . London: M. and S. Nealand, 1632.

Bronte, Charlotte. *Jane Eyre*. 1847: Toronto: Rinehart, 1958.

Bruce, Susan. *William Shakespeare: King Lear*. New York: Columbia University Press, 1998.

Buchanan, Roberdeau. *The Mathematical Theory of Eclipses*. Washington: Library of Congress, 1904.

Bullough, Geoffrey, ed. *Narrative and Dramatic Sources of Shakespeare*. London: Routledge and Kegan Paul, 1957–73.

Burgess, Anthony. *Shakespeare*. 1970. New York: Penguin, 1972.

Burton, Robert. *Anatomy of Melancholy*. London: 1621.

Byse, Fanny. *Milton on the Continent: A Key to L'Allegro and Il Penseroso*. Folcroft, PA.: The Folcroft P, 1903, 1969.

Casaubon, Meric, ed. *A True and Faithful relation of what passed for many years between Dr. John Dee and some spirits…* London: D. Maxwell, 1659.

Chapman, Allan. "Thomas Harriot: the first telescopic astronomer" *Journal of the British Astronomical Association* 118,6 (December 2008), 315–325; cf. Astronomy and Geophysics 50, 1 (2009), 1.27–1.33.

Christianson, John Robert. *On Tycho's Island: Tycho Brahe and his Assistants, 1570–1601*. Cambridge: Cambridge University Press, 2000.

Clark, Cumberland (1929) *Shakespeare and Science*. Honolulu: University Press of the Pacific, 2005.

Clark, Cumberland. *Astronomy in the Poets*. Folcroft, PA.: Folcroft Press, 1969.

Clark, David H., and F. Richard Stephenson. *The Historical Supernovae*. Oxford: Pergamon Press, 1977.

Dade, John. *An almanacke and prognostication in which this yeere of our Lord God...* 1605 ed. London, 1605.

Dade. *Almanacke 1605*. London, 1604.

Danby, John. *Shakespeare's Doctrine of Nature: A Study of King Lear*. London: Faber and Faber, 1948.

Davies, Sir John, *Orchestra, or A Poem of Dancing*, ed. Eustace M.W. Tillyard (London: Chatto and Windus, 1947) 27.

de Santillana, Giorgio. *The Crime of Galileo*. 1955. Chicago: University of Chicago Press, 1976.

Deacon, Richard. John Dee: *Scientist, Geographer, and Secret Agent to Elizabeth I*. London: Frederick Muller, 1968.

Dee, John. Asterismo,caelitus demissa ad orbem que Veneris: Iterumque in Caeli penetralia perpendiculariter retracta, post decimum fextum fuae apparitionis mensem." In A Letter, Nine years since, written and first published 1604.

Dee, John, ed. *The Elements of Geometrie of the most auncient Philosopher Euclide of Megara. With a very fruitfull Preface* by M. J. Dee. (London: H. Billingsley, 1570).

Dee, John. *De Stella admirada, in Cassiopeae Asterismo,caelitus demissa ad orbem que Veneris: Iterumque in Caeli penetralia perpendiculariter retracta, post decimum fextum fuae apparitionis mensem* Anno 1572. London: 1572.

Dee, John. *The Private Diary of John Dee.* 1581 August 26. http://www.gutenberg.org/etext/19553.

Devereux, E. J. "Review," *Renaissance Quarterly* 21,1 (spring, 1968), 89–91.

Digges, Thomas, *A Perfit Description of the Celestiall Orbes according to the most auncient doctrine of the Pythagoreans, lastely reuiued by Copernicus and by Geometrical Demonstrations approued.* London: 1576. Digges, Thomas, Pantometria (London: Henrie, 1571) 27. Digges, Thomas, Stratioticos (London: Henry Bynneman, 1579), 359.

Donne, John, "The First Anniversary," (lines 205–218) *The Complete Poetry of John Donne*, ed. J. T. Shawcross (Garden City: Anchor Books, 1967.) Drake, S., trans. Discoveries and Opinions of Galileo (New York: Anchor, 1957) 51 53, 93.

Drake, S. *Discoveries and Opinions of Galileo.* (New York: Doubeleday Anchor, 1957), 100.

Drake, S. *Essays on Galileo and the History and Philosophy of Science.* Vol. 1 (Toronto: University of Toronto Press, 1999).

Draper, J. W. "Shakespeare's Star-crossed Lovers." *Review of English Studies* 15 (1939): 20.

Dreyer, John L. E. *Tycho Brahe: a picture of scientific life and work in the sixteenth century.* Edinburgh, 1890.

Edberg, S. and Levy, D. *Observing Comets, Asteroids, Meteors, and the Zodiacal Light.* Cambridge: Cambridge University Press, 1994.

Eisler, Robert. *The Royal Art of Astrology.* London: Herbert Joseph Limited, 1946.

Ellis, Oliver. *Shakespeare as a scientist; his philosophical background.* Folcroft, PA.: Folcroft P, 1970. Espenak, Fred. Eclipse Home Page. <http://sunearth.gsfc.nasa.gov/eclipse/eclipse.html>.

Evans, Mary Anne (a.k.a. George Eliot), *Middlemarch.* 1871: Cambridge: Riverside Press, 1956.

Falconer, Alexander Frederick. *Shakespeare and the Sea.* New York: F. Ungar, 1964.

Fantoli, Annibale, and George V. Coyne. *Galileo: For Copernicanism and for the Church.* 2nd ed. Rome: Vatican Observatory Foundation, 1996.

Fitzgeoffrey, Charles. *Sir Francis Drake His honorable Life's commendation, and his Tragical Deathes Lamentation.* London: 1596.

Fleming, Abraham. *A Bright Burning Beacon, forewarning all wise Virgins to trim their Lamps against the Coming of the Bridegroom.* London: 1580.

Foakes, R. A., ed. *King Lear.* Erhardt: Arden Shakespeare, 1997.

Furness, Horace Howard, ed. *King Lear: The New Variorum Edition.* 1908. Toronto: Dover, 2000.

Garin, Eugenio, *Astrology in the Renaissance: The Zodiac of Life*. Trans. Carolyn Jackson, June Allen, and Clare Robertson. London: Routledge & Kegan Paul, 1983, 67.

Genest, J. *Some Account of the English Stage: From the Restoration in 1660 to 1830.* (Bath: Carrington, 1832),

Gingerich, Owen, "Great conjunctions, Tycho, and Shakespeare". *Sky & Telescope* (1981), 393–395.

Gottschalk, Paul A. "The Universe of Madness in King Lear." *Bucknell Review: A Scholarly Journal of Letters, Arts and Science* 19 (1971): 51–68.

Grafton, Richard. *Abridgement of the Chronicles of Englande, newly corrected and augmented to thys present yere of our Lord 1572*. London: Richard Tottyll, 1562, 1572.

Grant, Edward. *Planets, Stars, and Orbs: The Medieval Cosmos, 1200–1687.* Cambridge: Cambridge University Press, 1994.

Greenblatt, S. *Learning to Curse: Essays in Early Modern Culture*. London: Routledge, 1990.)

Greenblatt, S., Cohen, W., Howard, J. En, and Maus, K. E. *The Norton Shakespeare Based on the Oxford Edition*. New York: W.W. Norton, 2005.

Gresham, E. *Strange fearful and true newes, which hapned at Carlstadt, in the Kingdome of Croatia*. (London: G. Vincent and W. Blackwal, 1606.)

Grierson, Sir Robert. *The Poems of John Donne*. Ed. Vol. 1. London: Oxford University Press, 1933. 237–38.

Guillory, John. *Poetic Authority: Spenser, Milton, and literary history*. New York: Columbia University Press, 1983.

Hamilton, A. C., ed. *Edmund Spenser: The Faerie Queene*. London: Longman, 1977.

Harbage, A. "Dating Shakespeare's Sonnets." *Shakespeare Quarterly* 1, No. 2 (1950), 57–63.

Hariot, Thomas, *A Brief and True Report of the New Found Land of Virginia*. (London: 1588.

Harrison, G. B. *A Second Jacobean Journal*. Ann Arbor: University of Michigan P, 1958. Harrison, G. B., ed. *A Jacobean Journal, being a record of those things most talked about during the years 1603–1606*. London: George Routledge and Sons, 1941. Harrison, G. B., et al. "These Late Eclipses (and further letters)." *Times Literary Supplement* (London) 30 Nov. 1933: 856.

Harison, J. *The Summarye of the Chronicles of Englande Unto 1604*. (London:J. Harison,

Pick, John et al. *The Victorian Poets—A Guide to Research*. Cambridge, Mass.: Harvard University Press, 1969.

Shakespeare, W. *The Riverside Shakespeare*. Ed. Evans, G. B. Boston: Houghton Mifflin, 1974.

Virgil. *Virgil*. trans. Jackson, T. Oxford: Clarendon, 1921.

Wordsworth, W. *The Poetical Works of William Wordsworth*. London: Oxford University P 1604), 302, 324.

Hartung, J.B. "Was the formation of a 20-km-diameter crater on the Moon observed on June 18, 1178?" *Meteoritics* 11 (3), 1976, 187–194.

Harvey, John. *A discoursiue probleme concerning prohecies how far they are to be valued, or credited...* London: John Jackson, 1588.

Harvey, Richard. *An Astrological discourse upon the great and notable conjuction of the two superiour planets, Saturne & Jupiter, which happen the 28 day of April, 1583...* London: 1583.

Heywood, Thomas. *Pleasant dialogues and dramas...* London: R Oulton, 1637.

Hogg, Helen Sawyer, "Out of Old Books: The Introduction of the Copernican System to England", *Journal of the Royal Astronomical Society of Canada* 46 (1952), 195–201.

Ibsen, Lawrence, *Galsworthy: Naturalist Drama and Environmental Influences.* Ed. Catherine Cooper. Apr. 2001. 22 May 2004 <http://www.english literature.org/essays/ibsen.html>.

Jeayes, Isaac Herbert, ed. *Letters of Philip Gawdy 1579–1616*. London: J. B. Nicols and Sons, 1906.

Johnson, Francis R. *Astronomical Thought in Renaissance England.* 1941. New York: Octagon Books, 1968.

Johnson, Francis R. "The Influnce of Thomas Digges on the progress of modern astronomy in Sixteenth-Century England" *Osiris* 1 (January 1936), 390–410.

Johnson, Maureen. "'What's in a Name': Astrology and Onomastics in Romeo and Juliet." *Dissertation Abstracts International* 44 (1984): 3072A.

Jonson, Ben. *Volpone*. Dover ed. 1605; rpt. New York: Dover, 1994.

Joyce, James. *Ulysses*. 1922; New York: Penguin, 1968.

Kennedy, Edward. *Astronomy and Astrology in the Medieval Islamic World.* Hants, England: Adershot, 1998.

Kermode, Frank. "*King Lear*." *The Riverside Shakespeare*. Ed. G. Blakemore Evans. Boston: Houghton Mifflin, 1974.

Kirschbaum, Leo. *The Plays of Christopher Marlowe*. Cleveland: World Publishing, 1962.

Knight, G. Wilson. The Wheel of Fire: *Interpretations of Shakespearian Tragedy*. 1930. London: Methuen, 1972.

Kronk, Gary W. *Comets: A descriptive Catalog*. Hillside, N.J.: Enslow, 1984.
Kronk, Gary W. *Comets and Meteor Showers*. 08 June 2002. <http://comets.ams-meteors.org/meteors/showers/leonidancient.html>.

Lauaterus, Lewes. *Of Ghosts and Spirits Walking by Nyght*. London: 1572.
Levine, George. "Darwin and Pain: Why Science made Shakespeare nauseating." Raritan: *A Quarterly Review* 15 (1995): 97–114.

Levy, David H. *Starry Night: Astronomers and Poets Read the Sky*. Amherst: Prometheus Books, 2001.

Levy, David H. *Deep Sky Objects: The Best and Brightest from Four Decades of Comet Hunting*. Amherst, N.Y.: Prometheus Books, 2005.

Levy, David H. *Shoemaker by Levy: The Man who Made an Impact.* Princeton: Princeton University Press, 2000.)

MacCaffrey, Isabel K. Allegory and Pastoral in The Shepheardes Calender. 1969. *Essential Articles for the Study of Edmund Spenser*. Ed. A. C. Hamilton. Hamden, CT.: Archon Books, 1972. 549–68.

MacKenzie, Norman H., ed. *The Poetical Works of Gerard Manley Hopkins*. Oxford: Clarendon Press, 1990.

MacNeice, Louis. *Astrology*. Garden City, New York: Doubleday, 1964.

Marcus, J. and Olsen, M. "Biological Implications of Organic Compounds in Comets", in *Comets in the Post-Hally-Era*. ed. R. Newburn. Dordrecht: Kluwer Academic Publishers, 1991.

Marius, Simon, *Mundus Iovialis anno MDCIX Detectus Ope Perspicilli Belgici.* Die Welt des Jupiter, 1609 mit dem flämischen Teleskop entdeckt. Übersetzung; Hrsg. und bearb. von Joachim Schlör. Naturwiss. (von Alois Wilder), 1614.

Masson, David. *The Life of John Milton; narrated in connexion with the political, eccliastical and literary history of his time.* (Cambridge: Macmillan, 1859).

McAlinon, T. *Shakespeare's Tragic Cosmos*. Cambridge; Cambridge University Press, 1991.

Meadows, A. J. *The High Firmament: A Survey in Astronomy in English Literature*. Leicester: Leicester U P, 1969.

Meeus, Jean. *Mathematical Astronomy Morsels*. Richmond: Willmann-Bell, 1997.

Mercury Chaser's Calculator. Ed. John Walker. 2003. 21 July 2003 <http://www.fourmilab.ch/images/3planets/elongation.html>.

Miller, James L. *Measures of Wisdom : The Cosmic Dance in Classical and Christian antiquity*. Toronto: University of Toronto Press, 1986, 465.

Mixxe, Adze. Shakespeare and Astrology. 23 May 2002 <http://www.adze.com/zine/shakes.html>.

More, Henry. *Philosophicall Poems*. Cambridge: 1647.

Moss, Jean Ketz. *Novelties in the Heavens: Rhetoric and Science in the Copernican Controversy*. Chicago: Chicago University Press, 1993.

Nashe, Thomas. *The Unfortunate Traveller and Other Works* (London: Penguin, 1972).

Nausea, Friedrich. *A Treatise of Blazing Starres in Generall*. 1577: rpt. London: Bernard Alsop, 1618. Nicholl, Charles, A Cup of News: The Life of Thomas Nashe (Routledge and Kegan Paul, 1984) Nicolson, Marjorie Hope, "The Discovery of Space," Medieval and Renaissance Studies, Ed. O.B. Hardison, Jr. Chapel Hill: University of North Carolina Press, 1966.

Nicolson, Marjorie Hope. *The Breaking of the Circle: Studies in the Effect of the "New Science"_upon 17th Century Poetry*. rev. ed. Evanston: University of Chicago Press, 1960.

Nicolson, Marjorie. "Milton and the Telescope." ELH 2 (1935): 1–10. North, John David. The Fontana History of Astronomy and Cosmology. n.p., n.d. Nuttall, A.D. Shakespeare the Thinker. New Haven: Yale University Press, 2007.

Olson, Donald W., Marilyn. S. Olson, and Russell L. Doescher. "The Stars of Hamlet." *Sky & Telescope* Nov. 1998: 68–73.

Parr, Johnstone, and Alfred Harbage, eds. *William Shakespeare: The Complete Works. The Pelican Shakespeare* ed. New York: Viking P, 1969.

Patterson, W. B., Review of G. Curzon, Wotton and his Worlds: Spying, Science and Venetian Intrigues. *The Catholic Historical Review* Vol. 92 No. 4. (2006), 677–

679. Piecoski, Frank. Shakespeare's Astrology. 23 May 2002. <http://starcats.com/anima/shakespeare.html>.

Pingre, Alexandre Guy. *Annales Celestes du Dix-Septieme Siecle*. Paris: Gauthier-Villars, 1783, 1901.

Pinsky, Robert, personal communication, 20 March 2006.

Pinsky, Robert. *The Life of David*. New York: Schocken, 2005.

Rabkin, Norman. *Shakespeare and the Problem of Meaning*. Chicago: University of Chicago P, 1981.

Recorde, Roberte. *The Castle of Knowledge*. London: 1556.

Ribner, Irving. Notes on "The Life and Death of King John," *The Pelican Shakespeare* (New York: Viking Press, 1969, 1977) 623.

Reusche, Harry. "Edmund's Conception and Nativity in King Lear". *Shakespeare Quarterly* 20, No. 2 (Spring 1969) 161–164.

Robinson, L. J. "A Remarkable Eclipse of the Moon" *Sky & Telescope* 27 (March 1964) 142–146).

Rogers, Janine. "Lover, Poet, or Astronomer: Collecting Stars and Poems with David H. Levy" *Dalhousie Review* 89.2 (Summer 2009) 155–168.

Roman, Miriam. "L'art et la science de William Shakespeare: Des chiffres et des lettres." *Revue des Lettres Modernes: Histoire des Idees des Litteratures* (1999): 1431–36.

Roth, Cecil. *The Haggadah : A New Edition*. London:Soncino, 1934,

Rowlands, Samuel. *Humors ordinarie: where a man may be verie merrie, and exceeding vvell vsed for his sixe-pence*. London, 1605.

Russell, John L. *The Copernican System in Great Britain: The Reception of Copernicus' Heliocentric Theory*. Ed. Jerzy Dobrzycki. Dordrecht: D. Reidel, 1972.

Schoenbaum , Samuel. *William Shakespeare: A Compact Documentary Life*. Oxford: Oxford University Press, 1987.

Seneca, L. A. *Qeaestiones Naturales* VII, "De Cometis", XV11, 1.

Shakelton, Francis. *A blazyng Starre or Burning Beacon, Seene the 10. of October laste (and yet continewyng) set on fire by Gods providence, to call all sinners to earnest and speedie repentance*. (London: Kyngston for Henry Kirkham, 1580).

Shakespeare, W. *The Complete Works*. London: Collins, 1951, 1964. References to Shakespeare are from this edition.

Shapin, Steven. *A Social History of Truth: Civility and Science in Seventeenth Century England*. Chicago: University of Chicago Press, 1994.

Shapley, Harlow, *Beyond The Observatory* (New York: Scribner's, 1967), 141–150

Shoemaker, E. M., Hackman, R. J., and Eggleton, R. E. "Interplanetarey Correlation of Geologic Time." In *Advances in the Astronautical Sciences* Vol. 8. New York, Plenum Press, 70–89.

Smith, Logan Piersall. *The Life and Letters of Sir Henry Wotton*. Oxford: Clarendon, 1907.

Smith, W. D. "The Elizabethan Rejection of Judicial Astrology and Shakespeare's Practice" *Shakespeare Quarterly* 9 (1958): 159–76.

Sondheim, Moritz, "Shakespeare and the Astrology of his time." *Journal of the Warburg and Cortauld Institutes* 2, 243–259.

Spencer, Jeffry. *Heroic Nature: Ideal Landscape in English Poetry from Marvell to Thomson*. Chicago: Northwestern University Press, 1973.

Spevack, Marvin. *A Complete and Systematic Concordance to the Works of Shakespeare*. Hildesheim: Georg Olms, 1968.

Steane, J. B. Thomas Nashe: *The Unfortunate Traveler and other works*. London: Penguin, 1972.

Stoyan, Ronald. *Atlas of the Messier Objects: Highlights of the Deep Sky*. Cambridge, U.K.: Cambridge University Press, 2008.

Stubbs, W., ed. *Gervasii Cantuariensis Opera Historica: Chronica Gervasii, Rerum Britannicarum Medii Aevi Scriptores*, London, 1879, 73a.

The Galileo Project. Ed. Albert Van Helden. August 5, 1996. Rice University. 23 Mar. 2004 <http://es.rice.edu/ES/humsoc/Galileo/Images/Astro/Conceptions/tycho_univ.gif>

Thoren, Victor E. *The Lord of Uraniborg: A biography of Tycho Brahe*. Cambridge: Cambridge U P, 1990.

Tillyard, E. M. W. *The Elizabethan World Picture*. New York: Macmillan, 1944.

United Nations General Assembly, (announcement of International Year of Astronomy) 10863 (http://www.un.org/News/Press/docs/2007/ga10683.doc.htm).

Usher, Peter, "Galileo's Telescopy and Jupiter's Tablet" *Bulletin of the American Astronomical Society* 35:5 (2003) 1258.

Usher, Peter, "Shakespeare's Support for the New Astronomy" *The Oxfordian* 5 (2002) 132–146.

Wade, K., personal communication, 26 March 2006.

Weber, Alan Scott. *Shakespeare's Cosmology*. Diss. State U of New York, 1996.

Weis, René. *Shakespeare Unbound: Decoding a Hidden Life*. New York: Henry Holt, 2007.

Whitfield, Peter. *Astrology: A History*. New York: Harry N. Abrams, Inc., 2001.

Woolley, Benjamin. *The Queen's Conjurer: The Science and Magic of Dr. Dee*. London: HarperCollins, 2001.

Yeomans, Donald K. *Comets: A Chronological History of Observation, Science, Myth, and Folklore*. New York: John Wiley, 1991.

Works cited B: Chaps. 6–9.

1 Joseph Crehan, S. J. "Some Hopkins Memories", Hogkins Reserch Bulletin, 4 (1973), 29.

2 Geoffrey Chaucer, The Works of Geoffrey Chaucer, ed. F. N. Robinson (Boston: Houghton Mifflin, 1957). "The Miller's Tale", ff. 3458.

3 William Shakespeare, *The Riverside Shakespeare*, ed. G. Blakemore Evans (Boston: Houghton Mifflin, 1974), 1154.

4 John Donne, *The Poems of John Donne*, ed. Sir Herbert Grierson (London: Oxford University Press, 1933), "The First Anniversary", 11. 205–218.

5 Sir John Davies, *The Poems of Sir John Davies*, ed. R. Krueger (Oxford: Clarendon Press, 1975), "0rchestra", 103–104.

6 William Wordsworth, *The Poetical Works of William Wordsworth*, ed. T. Hutchinson, (London: Oxford University Press. 1917), "Star Gazers", 189.

Notes to Chap. 10

Hind, J.R. Letter, *London Times,* 1 August 1864, p. 5, col. e.

Guillemin, Amedée, *The World of Comets* (London: Sampson Low, 1877), p. 254f.

C. P. Olivier, C. P., Comets (Baltimore: Williams and Wilkins.1930). p. 155.

Dictionary of National Biograghy, 1917, vol. XV, p. 924. Also, H. House and G. Storey, ed., The Journals and Papers of Gerard Manley Hopkins, (London: Oxford University Press, 1959). pp. 419–420.

8 Guillemin, p. 322, fig, 55.

9 Coventry Patmore, *Poems* (London: George Bell, 1906), 255.

Notes to Chap. 11

1 W. H. Gardner, Gerard Manley Hopkins: A Study of Poetic Idiosyncracy in Relation to Poetic Diction (London: Oxford University Press, 1948); D. McChesney, *A Hopkins Commentary* (London: London Press, 1968), p. 57: P. Milward,_ *Commentary on the Sonnets of G. M. Hopkins* (Tokyo: HokuseidovPress, 1969). p.7; J. Pick, *Gerard Manley Hopkins: Priest and Poet* (London: Oxford University Press, 1966), p. 57; B. Bergonzi, Gerard Manley Hopkins (New York: Macmillan, 1977). p· 77.

2. Leslie C. Peltier, Starlight Nights: The Adventures of a Star Gazer (New York: Harper and Row, 1965), p. 41.

–

4 Peter Milward, A Commentary on the Sonnets of G. M. Hopkins (Tokyo: Hokuseido Press, 1969), p. 7.

5 J. Hillis Miller, "'Orion' in 'The Wreck of the Deutschland'", Modern Langtuage Notes, 76 (1961), 509–514.

6 Originally "Nec saevus Arcturi cadentls / impetus aut orientis Haedi" in *The Works of Horace.* ed. E. C. Wickwam (Oxford: Clarendon, 1877), p. 168; literal translation: "The Ruthless Onslaught of Arcturus / At setting, or the Kid at rising", in *The Odes of Horace: A Literal Translation*, trans. Arthur S. Way (London, Macmillan, 1936), p. 51.

7 For an excellent discussion of the mythological history of Orion, see Richard Hinckley Allen, *Star Names: Their Lore and Meaning* (New York: Dover, 1963), pp.303–320. Originally published in 1899 by Stechart as *Star—names and Their Megnings.*

8 John Keating, The Wreak of the Deutschland (Ohio: Kent State University, 1963), p. 73.

9 Sir Paul Harvey, *Oxford Companion to English Literature*, fourth ed. (Oxfords Clarendon, 1975), p. 494.

10 See Alfred Thomas, S. J. *Hopkins the Jesuit* (London: Oxford University Press. 1969). p. 188.

Notes to Appendix

g3 1 Nicholas De Vore, Encyclopaedia of Astrology (New York: Crown. 1947), pp. 136–154. Lists circumstances of all solar and lunar eclipses from about 1400 to 2107.

2 Virgil, Viggll, trans. John Jackson (Oxford: Clarendon, 1921), p. 30.

3 The planetary tables from 1850 to 1889 are found in *Die Deutsche Ephemeride* (Scherz Verlag: Aern-Munchen-Wien, Munich, 1974).

4 M James Glassford, trans. *Novum Organum Scientiarum* (Edinburgh: Edinburgh Printing Co., 1844 (MDCCCXLIV)), p. 37.

5 Guy Murchie, *Music of the Spheres: The Material Universe -- from Atom to Quasar, Simply Explained* (New York: Dover. 1967), frontispiece.

Primary Material

lA. Works by Hopkins (for reference codes see page 5U·

Poems of Gerard Manley Hopkins, fourth edn., ed. W. H. Gardner and N. H. Mackenzie. London: Oxford University Press, 1970.

The Journals and Papers of Gerard Manley Hopkins, ed. Humphrey House, completed Graham Storey. London: Oxford University Press, 1959.

The Sermons and Devotional Writings of Gerard Manley Hopkins, ed. Christopher Delvin, S. J. London: Oxford University Press, 1959.

The Letters of Gerard Manley Hopkins to Robert Bridges, ed. C. C. Abbott. London: Oxford University Press, 1955.

The Correspondence of Gerard Manley Hopkins and Richard Watson Dixon, ed. C. C. Abbott. London: Oxford University Press, 1955.

Further Letters of Gerard Manley Hopkins Including his Correspondence with Coventry Patmore, ed. C. C. Abbott. London: Oxford University Press, 1938; rev. and enlarged,1956. Both editions were consulted, although references in the text are all to the revised edition.

1B: Works by Other Poets

Bridces, Robert. *The Shoirter Poems of Robert Bridges* (Enlarged Edition). Oxford: Clarendon, 1931.

Dixon , R. W. *Poems by the late Dr. Richard Watson Dixon*, ed. Robert Bridges. London: Smith, Elder, 1909.

Patmore, Coventry. *Poems*. London: George Bell, 1906.

1C: Contemporary Journals and Newspapers

Astronomical Notices. Royal Astronomical Society, 1858.

Cornhill. 1861-1864. Especially "The Comet" Oct. 1862, pp.550–551.

Illustrated London News. May through October, 1858 (Donati's Comet); July and August, 1861 (Tebutt's Comet). 16 Aug. 1862 (Rosa's Comet, actually 109P/ Swift-Tuttle , p. 179.

London Review. 1861. Articles on Tebutt's Comet (C/1861 J1):
6 July 1861, pp. 15–18.
13 July 1861, pp. 44–47.
20 July 1861, p. 76.
20 July 1861, pp. 89–90.
27 July 1861, pp. 108–109.
3 August 1861, pp. 137–140.
10 August 1861, pp. 181–182.
24 August 1861, p. 237.
14 September 1861, p. 330.

Times (London). Articles on Tebutt's Comet (C/1861 J1):
1 July 1861, p. 5–
2 July 1861, p. 5.
3 July 1861. P. 5.
4 July 1861, p. 12.
6 July 1861, p. 12.
9 July 1861, p. 9.
15 July 1861, p. 10.
Article on Tempel's Comet (C/1864 N1):
1 August 1864, p. 5.

Nature
Whipple, G. M. Letter 13 March 1884, p. 453.
Blackhouse, T. W. Letter 31 (1884), 28, p. 28.

Secondary Material

2A: Works on Hopkins

Ball, Patricia M. *The Science of Aspects: The Changing Role of Fact in the Works of Coleridge, Ruskin and Hopkins*. London: Athlone Press, 1971.

Bender, Todd K. Gerard Manley Hopkins: *The Classical Background and Critical Recreption of his Work*. Baltimore: Johns Hopkins University Press, 1966.

Berronzi, B. *Gerard Manley Hopkins*. New York: Macmillan, 1977.

Cotter, James Finn. *Inscapes: The Christology and Poetry of Gerard Manley Hopkins*. Pittsburgh: University of Pittsburgh Press, 1972.

Crehan, Joseph, S. J. "Some Hopkins Memories"· *The Hopkins Research Bulletin*, 4 (1973): 29.

Dilligan, Robert J. and Bender, Todd K. *A Concordance to the English Poetry of Gerard Manley Hopkins.*· University of Wisconsin Press, 1970.

Dunne, Tom. Gerard Manley Hopkins: A Comprehensive Bibliography. Oxford: Clarendon Press, 1976.

Gardner, William Henry. *Gerard Manley Hopkins: A Study of Poetic Idiosyncracy in Relation to Poetic Diction.* New Haven: Yale University Press, 1948–49.

Keating, John Edward. *The Wreck of the Deutschland: An Essay and Commentary.* Kent, Ohio: Kent State University Press, 1963.

Kenyon Critics. *Gerard Manley Hopkins.* New York: Dennis Dobson, 1945.

McChesney, Donald. *A Hopkins Commentary—an Explanatory Commentary on the Main Poems, 1875-1889.* London: London Press, 1968.

MacKenzie, Norman H. *Hopkins* (writers and Critics). Edinburgh: Oliver and Boyd, 1968.

MacKenzie, Norman H. "Gerard Manley Hopkins' Spelt from Sybil's Leaves". Malahat Review 26 (1973), pp. 218–228.

MacKenzie, Norman H. "Hopkins Among the Victorians: Form in Art and Nature". English Studies Today, Third Series. Edinburgh: Edinburgh University Press, 1964, pp. 155-168,

Miller, J. H. "0rion in 'The Wreck of the Deutschland'", *Modern Language Notes* 76 (1961), pp. 509–514.

Milward, Peter. *A Commentary on the Sonnets of G. M. Hopkins.* Tokyo: Hoksuido Press, 1969.

Pick, John (ed.) *The Windhover* (Merill Literary Casebook series.) Columbus: Merill, 1969.

Ritz, Jean-Georges. *Robert Bridges and Gerard Hopkins, 1863–1889—A Literary Friendship.* London: Oxford University Press, 1960.

Ritz, Jean-Georges. *Le Poète Gērard Manley Hopkins, 1844-1889: L'Homme et L'Oeuvre.* Paris: Didler, 1963.

Ruggles, Eleanor. *Gerard Manley Hopkins, A Life.* New York: Norton, 1944.

Schneider, Elizabeth W. *The Dragon in the Gate: Studies in the Poetry of G. M. Hopkins.* Berkeley and Los Angeles: University of California Press, 1968. See also 8.I(1981), 4–8.

Thomas. A. *Hopkins the Jesuit: the Years of Training.* London: Oxford University Press, 1969.

Weyamd. Norman (ed.). Immortal Diamands: Studies in Gerard Manley Hopkins. New York: Shaed and Ward. 1949.

2B:Astronomical and Related Works

Allen, D.C. *The Star-Crossed Renaissance: the Quarrel about Astrology and Its Influence in England. Durham, N. C.: 1941.*

Allen, Richard H. *Star Names: Their Lore and Meaning*. 1899: rpt. New York: Dover, 1963.

Bacon, Francis. *Novum Organum Scientriarum.* Trans. Glassfordf, J. Edinburgh: Edinburgh Printing Co.:, 1844.

Ball, Sir Robert S. The_Story_of the Heavens. London: Cassell & Co., 1886.

Bennett, H. S. "Science and Information in English Writings of the 15th Century". Modern Language Review, 39 (1944), pp.1–8.

Berry, Arthur. *A Short History of Astronomy*. 1898; rpt. New York: Dover, 1961.

Bush, Douglas. *Science and English Poetry: A Historical Sketch, 1590–1950.* New York: Oxford University Press, 1950.

Crawford, Russell, T. *Determination of orbits of Comets and Asteroids*. New York: McGraw-Hill, 1930.

Croker, J. W., "Poems by Alfred Tennyson" *Quarterly Review*, Spring 1833.

De Santillana, Giorgio. *The Crime of Galileo*. 1955: rpt. Chicago: University of Chicago Press (Midway), 1976.

De Vore, Nicholas. *Encyclopaedia of Astrology*. New York: Crown, 1947.

Flammarion, Camille. *The Atmosphere*. Ed. Glaisher, J. New York: Harper and Bros. n.d. (1872?)

Guillemin, Amedee. *The World of Comets*. Trans. and ed. Glaisher, James. London: Sampson, Low, et al., 1877.

Harris, V. *All Coherence Gone*. Chicago: U. of Chicago Press, 1949.

Hind- J· Russell. *The Comet of 1556, being Popular Replies to Every-Day Questions referring to its antiucipated reappearanbce with some observations on the apprehension of danger from Comets.* . London: John W. Parker, 1857.

Hind- J· Russell. *The Comets: A Descriptive Treatise Upon those Bodies*. London: Johgn W. Parker, 1852.

Johnson, Francis R. *Astronomical Thought in Renaissance England*. New York: Octagon Books, 1968.

Jones, R. F. "The Rhetoric of Science in England of the mid-17th Century". In *Restoration and Eighteenth Century Literature: Essays in honour of A. D. McKillop*. Chicago, 1963.

Kocher, Paul H. *Science and Religion in Elizabethan England.* San Marino: Huntingdon Library, 1953.

Levy, David H. *The Man Who Sold the Milky Way: A biography of Bart Bok*) Tucson: University of Arizona Press, 1993), 64.

Lockyer, J. Norman. *The Dawn of Astronomy: A Study of the Temple Worship ansd Mythology of the Ancient Egyptians.* Cassell and Co., 1894; rpt. Cambridge,. Mass.: MIT Press, 1964.

Lowell, Percival. *Mars and its* Canals. New York: Macmillan, 1906.

Maynard, K. "Science in Early English Literature, 1550-1650". *Isis*, 17 (1932).

Murchie, G. *Music of the Sphere: The Material Universe from Atom to Quasar, Simply Explained*. New York: Dover, 1967.

Nicholson, Marjorie Hope. *The Breaking of the Circle: Studies in the Effect of the "New Science" upon 17th Century Poetry.* Evanston: University of Chicago Press, 1950: rev. New York: 1960.

Nicholson, Marjorie Hope. *Science and Imagination.* Ithaca: Great Seal Books, 1956.

Nicholson, Marjorie Hope. "The Discovery of Space". *Medieval and Renaissance Studies.* Ed. 0. B. Hardison, Jr. Chapel Hill}: Chapel Hill: University of North Carolina Press, 1966.

Norton, Arthur P. and Inglis. *Norton's Star Atlas and Telescopic Handbook.* 14th ed. Edinburgh: Gall and Inglis, 1959.

Oliver, Charles P. *Comets.* Baltimore: Williams and Wilkins, 1930.

Peltier, Leslie C. *Starlight Nights: The Adventures of a Star-Gazer.* New York: Harper and Row, 1965).

Sarton, George. *The Appreciation of Ancient and Medieval Science during the Renaissance (1450–1600).* Philadelphia: University of Pennsylvania Press, 1955.

Sarton, George. *Ancient Science and Modern Civilization.* Lincoln: University of Nebraska Press, 1954.

Sky & Telescope, 23, No. 2 (1962), 92–93.

Verlag, Scherz. *Die Deutsche Ephemeride.* Bern-Mönchen-Wien: Otto Wilhelm Barth-Verlag, 1974.

Webb, Rev. T. . Celestial Objects for Common Telescopes. 1917: rpt. ed. Mayall, M. New York: Dover, 1962.

2C: Other Works

Butler, G. F., "Milton's Meeting with Galileo: A Reconsideration", *Milton Qurterly* 39 (3), 2005, 132–139.

Chaucer, Geoffrey. *The Works of Geoffrey Chaucer*. Ed. Robinson, F. N. Boston: Houghton Mifflin, 1957.

Donoghue, D. *The Ordinary Universe*. London: Faber, 1968.

Donne, John. *The Poems of John Donne*. Ed. S. Grierson, Sir H. London: Oxford University Press, 1933.

Davies, Sir John. *The Poems of Sir John Davies*. Ed. Krueger, R. Oxford: Clarendon Press, 1975.

Dreyer, J. L. E., ed. *The Scientific Papers of Sir William Herschel* (London: The Royal Society and The Royal Astronomical Society, 1912) 2: 30–38.

Gibson, Walker. "Behind the Veil: a Distinction Between Poetic and Scientific Language in Tennyson, Lyell and Darwin". *Victorian Studies* II. Sept. 1958, pp. 60–68.

Harvey, Sir Paul. *Oxford Companion to English Literature*. Fourth ed. Oxford: Clarendon, 1975.

Hoyt, W. G., *Planets X and Pluto* (Tucson: University of Arizona Press, 1980), 12.

Jones, K. G., *Messier's Nebulae and Star Clusters,* 2nd ed. (London: Cambridge University Press, 1991), 347.

Lewis, C. S. "The Last Battle" (1956: New York: HarperCollins, 1984), 189.

Lubbock, C. *The Herschel Chronicle: The Life-Story of William Herschel and his sister, Caroline Herschel, edited by his granddaughter Constance A. Lubbock.* (New York: MacMillan, 1933), 60.

Ross, R. H., ed. *In Memoriam* (New York: Norton, 1973), 104. Cf. Sir Charles Tennyson's essay *In Memoriam*, 105.

Sheehan, W. *Worlds in the Sky: Planetary Discovery from Earliest Times Through Voyager and Magellan* (Tucson: University of Arizona Press, 1992), 132.

Bibliography

Primary Material—1A. Works by Hopkins (for reference codes see page 5U)

Poems of Gerard Manley Hopkins, fourth edn., ed. W. H. Gardner and N. H. Mackenzie. London: Oxford University Press, 1970.

The Journals and Papers of Gerard Manley Hopkins, ed. Humphrey House, completed Graham Storey. London: Oxford University Press, 1959.

The Sermons and Devotional Writings of Gerard Manley Hopkins, ed. Christopher Delvin, S. J. London: Oxford University Press, 1959.

The Letters of Gerard Manley Hopkins to Robert Bridges, ed. C. C. Abbott. London: Oxford University Press, 1955.

The Correspondence of Gerard Manley Hopkins and Richard Watson Dixon, ed. C. C. Abbott. London: Oxford University Press, 1955.

Further Letters of Gerard Manley Hopkins Including his Correspondence with Coventry Patmore, ed. C. C. Abbott. London: Oxford University Press, 1938; rev. and enlarged, 1956. Both editions were consulted, although references in the text are all to the revised edition.

Primary Material—1B: Works by Other Poets

Bridges, Robert. *The Shorter Poems of Robert Bridges* (Enlarged Edition). Oxford: Clarendon, 1931.

Dixon , R. W. *Poems by the late Dr. Richard Watson Dixon*, ed. Robert Bridges. London: Smith, Elder, 1909.

Greenblatt, S., *et al. The Norton Shakespeare* (1997: 2nd ed. New York and London: W. W. Norton, 2008) 442.

Morris, W. "Summer Dawn", in *The Oxford Book of English Verse: 1250–1900.* (Oxford: Oxford University Press, 1900).

Patmore, Coventry. Poems. London: George Bell, 1906.

D.H. Levy, *The Starlight Night*, Astrophysics and Space Science Library 419,
DOI 10.1007/978-3-319-19878-1

Primary Material—1C: Contemporary Journals and Newspapers

Astronomical Notices. Royal Astronomical Society, 1858.
Cornhill. 1861–1864. Especially “The Comet” Oct. 1862, pp. 550–551.
Illustrated London News. May through October, 1858 (Donati’s Comet); July and August, 1861 (Tebutt's Comet). 16 Aug. 1862 (Rosa’ Comet, actually 109P/Swift-Tuttle), p. 179.
London Review. 1861. Articles on Tebutt's Comet (C/1861 J1):
6 July 1861, pp. 15–18.
13 July 1861, pp. 44–47.
20 July 1861, p. 76.
20 July 1861, pp. 89–90.
27 July 1861, pp. 108–109.
3 August 1861, pp. 137–140.
10 August 1861, pp. 181–182.
24 August 1861, p. 237.
14 September 1861, p. 330.
Times (London). Articles on Tebutt’s Comet (C/1861 J1):
1 July 1861, p. 5.
2 July 1861, p. 5.
3 July 1861. p. 5.
4 July 1861, p. 12.
6 July 1861, p. 12.
9 July 1861, p. 9.
15 July 1861, p. 10.
Article on Tempel's Comet (C/1864 N1):
1 August 1864, p. 5.
Nature
Whipple, G. M. Letter 13 March 1884, p. 453.
Blackhouse, T. W. Letter 31 (1884), 28, p. 28.

Secondary Material—2A: Works on Hopkins

Ball, Patricia M. *The Science of Aspects: The Changing Role of Fact in the Works of Coleridge, Ruskin and Hopkins*. London: Athlone Press, 1971.
Bender, Todd K. Gerard Manley Hopkins: *The Classical Background and Critical Reception of his Work*. Baltimore: Johns Hopkins University Press, 1966.
Berronzi, B. *Gerard Manley Hopkins*. New York: Macmillan, 1977.
Cotter, James Finn. *Inscapes: The Christology and Poetry of Gerard Manley Hopkins*. Pittsburgh: University of Pittsburgh Press, 1972.
Crehan, Joseph, S. J. “Some Hopkins Memories”. *The Hopkins Research Bulletin*, 4 (1973): 29.
Dilligan, Robert J. and Bender, Todd K. *A Concordance to the English Poetry of Gerard Manley Hopkins*. University of Wisconsin Press, 1970.
Dunne, Tom. Gerard Manley Hopkins: A Comprehensive Bibliography. Oxford: Clarendon Press, 1976.
Gardner, William Henry. *Gerard Manley Hopkins: A Study of Poetic Idiosyncrasy in Relation to Poetic Diction*. New Haven: Yale University Press, 1948–49.
Keating, John Edward. *The Wreck of the Deutschland: An Essay and Commentary*. Kent, Ohio: Kent State University Press, 1963.

Kenyon Critics. *Gerard Manley Hopkins*. New York: Dennis Dobson, 1945.
McChesney, Donald. *A Hopkins Commentary—an Explanatory Commentary on the Main Poems, 1875–1889*. London: London Press, 1968.
MacKenzie, Norman H. *Hopkins* (writers and Critics). Edinburgh: Oliver and Boyd, 1968.
MacKenzie, Norman H. "Gerard Manley Hopkins' Spelt from Sybil's Leaves". Malahat Review 26 (1973), pp. 218–228.
MacKenzie, Norman H. "Hopkins Among the Victorians: Form in Art and Nature". English Studies Today, Third Series. Edinburgh: Edinburgh University Press, 1964, pp. 155–168.
Miller, J. H. "Orion in 'The Wreck of the Deutschland'", *Modern Language Notes* 76 (1961), pp. 509–514.
Milward, Peter. *A Commentary on the Sonnets of G. M. Hopkins*. Tokyo: Hokuseido Press, 1969.
Pick, John (ed.) *The Windhover* (Merill Literary Casebook series). Columbus: Merill, 1969.
Ritz, Jean-Georges. *Robert Bridges and Gerard Hopkins, 1863–1889—A Literary Friendship*. London: Oxford University Press, 1960.
Ritz, Jean-Georges. *Le Poète Gērard Manley Hopkins, 1844–1889: L'Homme et L'Oeuvre*. Paris: Didier, 1963.
Ruggles, Eleanor. *Gerard Manley Hopkins, A Life*. New York: Norton, 1944.
Schneider, Elizabeth W. *The Dragon in the Gate: Studies in the Poetry of G. M. Hopkins*. Berkeley: University of California Press, 1968.
Thomas. A. *Hopkins the Jesuit: the Years of Training*. London: Oxford University Press, 1969.
Weyamd. Norman (ed.). Immortal Diamond: Studies in Gerard Manley Hopkins. New York: Sheed and Ward. 1949.

Secondary Material—2B: Astronomical and Related Works

Allen, D.C. *The Star-Crossed Renaissance: the Quarrel about Astrology and Its Influence in England. Durham, N. C.: 1941.*
Allen, Richard H. *Star Names: Their Lore and Meaning*. 1899: rpt. New York: Dover, 1963.
Bacon, Francis. *Novum Organum Scientriarum.* Trans. Glassford, J. Edinburgh: Edinburgh Printing Co.: 1844.
Ball, Sir Robert S. The Story of the Heavens. London: Cassell & Co., 1886.
Bennett, H. S. "Science and Information in English Writings of the 15th Century". Modern Language Review, 39 (1944), pp. 1–8.
Berry, Arthur. *A Short History of Astronomy*. 1898; rpt. New York: Dover, 1961.
Bush, Douglas. *Science and English Poetry: A Historical Sketch, 1590–1950.* New York: Oxford University Press, 1950.
Crawford, Russell, T. *Determination of orbits of Comets and Asteroids*. New York: McGraw-Hill, 1930.
De Santillana, Giorgio. *The Crime of Galileo*. 1955: rpt. Chicago: University of Chicago Press (Midway), 1976.
De Vore, Nicholas. *Encyclopaedia of Astrology*. New York: Crown, 1947.
Flammarion, Camille. *The Atmosphere*. Ed. Glaisher, J. New York: Harper and Bros. n.d. (1872?).
Guillemin, Amedee. *The World of Comets*. Trans. and ed. Glaisher, James. (London: Sampson, Low, et al., 1877).
Harris, V. *All Coherence Gone*. Chicago: U. of Chicago Press, 1949.
Hind, J. Russell. *The Comet of 1556, being Popular Replies to Every-Day Questions referring to its anticipated reappearance with some observations on the apprehension of danger from Comets* (London: John W. Parker, 1857).
Hind, J. Russell. *The Comets: A Descriptive Treatise Upon those Bodies*. London: John W. Parker, 1852.

Johnson, Francis R. *Astronomical Thought in Renaissance England.* New York: Octagon Books, 1968.
Jones, R. F. "The Rhetoric of Science in England of the mid-17th Century". In *Restoration and Eighteenth Century Literature: Essays in honour of A. D. McKillop*. Chicago, 1963.
Kocher, Paul H. *Science and Religion in Elizabethan England.* San Marino: Huntingdon Library, 1953.
Levy, D. *David Levy's Guide to Observing Meteor Showers* (Cambridge: Cambridge University Press, 2008), 12.
Lockyer, J. Norman. *The Dawn of Astronomy: A Study of the Temple Worship and Mythology of the Ancient Egyptians.* Cassell and Co., 1894; rpt. Cambridge, Mass.: MIT Press, 1964.
Lowell, Percival. *Mars and its Canals*. New York: MacMillan, 1906.
Marsden, B. G. "The Sungrazing Comet Group", *Astronomical Journal* 72 (1967) 1179.
Maynard, K. "Science in Early English Literature, 1550–1650". *Isis*, 17 (1932).
Millhauser, M. *Fire and Ice: The Influence of Science in Tennyson's Poetry* (Lincoln: The Tennyson Society, Tennyson Research Center, 1971), 19.
Murchie, G. *Music of the Sphere: The Material Universe from Atom to Quasar, Simply Explained.* New York: Dover, 1967.
Nicholson, Marjorie Hope. *The Breaking of the Circle: Studies in the Effect of the "New Science" upon 17th Century Poetry.* Evanston: University of Chicago Press, 1950: rev. New York: 1960.
Nicholson, Marjorie Hope. *Science and Imagination.* Ithaca: Great Seal Books, 1956.
Nicholson, Marjorie Hope. "The Discovery of Space". *Medieval and Renaissance Studies*. Ed. 0. B. Hardison, Jr. Chapel Hill: University of North Carolina Press, 1966.
Norton, Arthur P. and Inglis. *Norton's Star Atlas and Telescopic Handbook.* 14th ed. Edinburgh: Gall and Inglis, 1959.
Oliver, Charles P. *Comets*. Baltimore: Williams and Wilkins, 1930.
Sarton, George. *The Appreciation of Ancient and Medieval Science during the Renaissance (1450–1600)*. Philadelphia: University of Pennsylvania Press, 1955.
Sarton, George. *Ancient Science and Modern Civilization.* Lincoln: University of Nebraska Press, 1954.
Seargent, D. *The Greatest Comets in History: Broom Stars and Celestial Scimitars* (New York: Springer, 2009), 99.
Sky & Telescope, 23, No. 2 (1962), 92–93.
Verlag, Scherz. *Die Deutsche Ephemeride*. Bern-München-Wien: Otto Wilhelm Barth-Verlag, 1974.
Webb, Rev. T. Celestial Objects for Common Telescopes. 1917: rpt. ed. Mayall, M. New York: Dover, 1962.

Secondary Material—2C: Other Works

Ashbrook, J., "Tycho Brahe's Nose", *Sky & Telescope*, June 1965, 353–354. See also Ashbrook's *The Astronomical Scrapbook* (Cambridge, Mass: Cambridge University Press, and Cambridge, England: Cambridge University Press, 1984) 3–5.
Chaucer, Geoffrey. *The Works of Geoffrey Chaucer*. Ed. Robinson, F. N. Boston: Houghton Mifflin, 1957.
Cotter, J. F. *Inscape: The Christology and Poetry of Gerard Manley Hopkins* (Pittsburgh: University of Pittsburgh, 1972), 172.
Donoghue, D. *The Ordinary Universe*. London: Faber, 1968.
Donne, John. *The Poems of John Donne*. Ed. S. Grierson, Sir H. London: Oxford University Press, 1933.
Davies, Sir John. *The Poems of Sir John Davies*. Ed. Krueger, R. Oxford: Clarendon Press, 1975.

Gibson, Walker. "Behind the Veil: a Distinction Between Poetic and Scientific Language in Tennyson, Lyell and Darwin". *Victorian Studies* II. Sept. 1958, pp. 60–68.

Harvey, Sir Paul. Oxford Companion to English Literature. Fourth ed. Oxford: Clarendon, 1975.

Miller, J. H. "'Orion' in 'The Wreck of the Deutschland'". *Modern Language Notes* 76 (1961), 509–514.

Peltier, Leslie C. Starlight Nights: The Adventures of a Star-Gazer. New York: Harper and Row, 1965.

Pick, John et al. *The Victorian Poets—A Guide to Research*. Cambridge, Mass.: Harvard University Press, 1969.

Shakespeare, W. *The Riverside Shakespeare*. Ed. Evans, G. B. Boston: Houghton Mifflin, 1974.

Virgil. Virgil. trans. Jackson, T. Oxford: Clarendon, 1921.

Wordsworth, W. *The Poetical Works of William Wordsworth*. London: Oxford University Press.

Wickwam, E.C., ed. *The Works of Horace* (Oxford: Clarendon Press, 1877), 168. Originally "Nec saevus Arcturi cadentis/Impetus aut orientis Haeedi"; literal translation: "The ruthless onslaught of Arcturus/At setting, or the Kid at rising" in *The Odes of Horace: a Literal Translation,* trans. Arthur S. Way (London: Macmillan, 1936), 51.

Index

D.H. Levy, *The Starlight Night*, Astrophysics and Space Science Library 419,
DOI 10.1007/978-3-319-19878-1

MIX
Papier aus verantwortungsvollen Quellen
Paper from responsible sources
FSC® C105338

If you have any concerns about our products,
you can contact us on
ProductSafety@springernature.com

In case Publisher is established outside the EU,
the EU authorized representative is:
Springer Nature Customer Service Center GmbH
Europaplatz 3, 69115 Heidelberg, Germany

Printed by Libri Plureos GmbH
in Hamburg, Germany